Hearing Happiness

Hearing Happiness

Deafness Cures in History

JAIPREET VIRDI

The University of Chicago Press

Chicago and London

The University of Chicago Press, Chicago 60637
The University of Chicago Press, Ltd., London
© 2020 by The University of Chicago
Published 2020
Printed in the United States of America

29 28 27 26 25 24 23 22 21 20   1 2 3 4 5

ISBN-13: 978-0-226-69061-2 (cloth)
ISBN-13: 978-0-226-69075-9 (e-book)
DOI: https://doi.org/10.7208/chicago/9780226690759.001.0001

Published with support of the Susan E. Abrams Fund.

Library of Congress Cataloging-in-Publication Data

Names: Virdi, Jaipreet, author.
Title: Hearing happiness : deafness cures in history / Jaipreet Virdi.
Description: Chicago ; London : University of Chicago Press, 2020. |
    Includes bibliographical references and index.
Identifiers: LCCN 2019041405 | ISBN 9780226690612 (cloth) |
    ISBN 9780226690759 (ebook)
Subjects: LCSH: Deafness—Treatment—United States—History—19th
    century. | Deafness—Treatment—United States—History—20th
    century. | Quacks and quackery—United States. | Medicine—
    United States—History.
Classification: LCC RF291 .V57 2020 | DDC 362.4/2—dc23
LC record available at https://lccn.loc.gov/2019041405

To Geoffrey.
This is yours as much as it is mine.

None of the medicines that are advertised to the public as "deafness cure" will cure deafness, nor will a good many of the hearing devices that are advertised.

Contents

Illustrations

Preface

My friend Erin and I attended the same "special" class for hearing-impaired children and became friends when we were in grade six. We lost touch and drifted apart some time at the end of our high school years, partly because we were at different schools and partly because our lives were heading down different paths. For most of our lives, Erin and I were told we were "hard-of-hearing" though we were both audiologically classified as profoundly deaf. As such, we grew up experiencing struggles of identity formation. Erin found a sense of belonging in Deaf culture. I struggled to "pass" as hearing and finally asserted my deafness in my thirties after continuous difficulties transitioning from analog to digital hearing aids.

Perhaps it was serendipity, but not long after my assertion, Erin and I realized we had both settled in the same city, a few blocks from each other. We had kept in touch over the years, but it was in this city where our friendship blossomed. Because I was working on this book, many of our conversations centered on the meanings of deafness and identity through history—on hereditary deafness and the impact it has on families, on cochlear implants, on community and belonging, and on historical definitions of deafness.

Erin reminded me that discussions about hearing being "restored" after being "lost" do not fit with the discourse of deafness and Deaf culture, because most Deaf people were born deaf and thus, never had any hearing to "lose." I should clarify then, that this is a book about oralist experiences of deafness, hardness of hearing, and hearing impairment, not cultural Deafness, though that is not to say that the Deaf do not figure in the story. My aim is not to deny the richness of Deaf culture, but to enrich our understanding of a nuanced and considerably underrepresented aspect of deaf history: that of the medical and technological avenues for "curing" hearing loss. Most of the actors in this book are those we would classify as "hard of hearing" and largely relied on oral communication. That is not to say that these actors themselves may have, or have not, considered themselves as deaf or Deaf. I've adhered to the historical record as faithfully as possible, so I use terms such as "deaf and dumb," "deaf-mutes," "hearing impaired," or "deafened" as people at the time used them.

The contents of this book should not be construed as medical advice. I am a historian of medicine, not a physician. My object is to contextualize and deepen your experiences of medicine, health, and disability. If you are concerned about your hearing, visit a qualified specialist for an assessment.

Cures of Yesterday

Another experience that people who are gradually deaf go through is that they think they are going to get better. They can't believe it is going to happen to them. There must be a cure somewhere, in this modern age of miracles. It is of course, on the surface a shame to take the hope away from them.

Logan Clendening, 1943[1]

The belief in cures tethers us not only to what we remember of our embodied selves in the past but also to what we hope for them in the future. And when those hopes are predicated on cure technology not yet invented, our body-minds easily become fantasies and projections.

Eli Clare, 2017[2]

When I was four years old, I became ill with bacterial meningitis and nearly died.

It was November 1985. We were living in Kuwait at the time, before the Gulf War displaced and scattered my close-knit family across disparate lands. These were still days marked by weekend treks to the desert, where my dad cooked kebabs; days of unexpected playdates and jubilant birthdays. I was an energetic, talkative child who learned six languages but remained shy when reciting the Sikh *ik onkar* prayer from memory in front of our family and congregation. There's a photo of me in the family album: I'm sitting in a bumper car at a fair, wearing a gray dress with a white collar, my hair in plaits, my cheeks flushed, and my eyes forlorn. I never thought it was a happy photo and often wondered why it was taken in the first place. Other times I'm glad it remains secured in the album because it serves as a memento of the time everything changed. It captures the moment when my parents realized I was seriously ill.

The doctor initially dismissed my parents' concerns about my febrile state. It was nothing more than a bout of the flu, he repeatedly soothed them. Yet my symptoms refused to cease, and my temperature spiked to dangerously high levels. Shortly thereafter, I was hospitalized. Through blurry memories, I recall the screams and the tears. The scissors cutting my favorite salwar kameez, the burgundy one with embroidered white roses on soft silk. I cherished it. It was the most beautiful outfit I owned and within a few seconds, the doctor slashed it from my body. All the beauty and love it once embodied were reduced to shredded fabric causally thrown on the floor. The scissors remained on the bed next to my frail, thin body. I sat there in my white underclothes, crying, unable to comprehend what had happened.

I remained bedridden for six months. I recall the loneliness and confusion: shadows dancing on the walls, wails as my mom closed the door behind her as she left for the evening, the feverish heat that stole many hours of daylight. The friendly voice of a neighboring boy who kept me awake with grand stories of everlasting friendship, stories whose magic and mystery floated through darkness, dissipating in the fluorescent streams that emitted from the hallway. For all I know, the boy never existed, his voice nothing more than mist in the air, carried away by the ghosts of the night. There are moments that stay with you, dividing the before and the after as silence slowly cloaks your world. The face of loved ones who look at you with pity, the sighs of exhaustion as they patiently try to understand your jumbled voice. The soft sobs that cradle you to sleep, the unknown sounds resonating through the hospital hallways.

For a long time, I never understood that I lost my hearing. I believed the world had changed and everyone simply stopped noticing me.

Like many parents faced with the illness of a child, my mother was overjoyed about my return to health but devastated by the realization that I would never be the same. I was deaf. My speech was reduced to a series of incomprehensible garbles. Convinced my deafness was divine punishment for her failings as a parent, my mother worried endlessly about how she could bring back that boisterous child of hers. How to explain to me that the sickness that ravaged through my brain had transformed both of our worlds. She fretted over communication, because she no longer knew how to speak to a child who could not—or would not—respond, a child who showed no indication of understanding simple instructions, let alone complex thoughts. She demanded the doctor correct his error and fix what was broken. She asked everyone for advice and suggestions for medical remedies that could possibly heal me, no matter how farfetched they seemed.

Countless hours were spent in the gurdwara (Sikh temple) asking for divine intervention. When the gurdwaras of Kuwait proved insufficient, I was whisked away to the Golden Temple in India. My grandfather recommended a series of healing herbs and tinctures. Sympathetic family and friends prescribed their own herbal solutions, secret recipes passed through family hands for generations. Once, when a Hindu shaman was traveling through town, I was taken to visit him in the hopes his direct link to divinity could be passed on to me. I was instructed to sit next to him as he spoke to the congregation; much to my mother's disapproval, I squirmed throughout the sermon, trying to remove the heavy hand placed upon my head. The shaman then presented me with a copper bracelet blessed with healing powers, which I wore until the circumference of my wrist outgrew that of the bracelet. They bejeweled me with rings they said were endowed with natural powers, burned special incense for me, fed me healing foods. While my family fished for a miracle, I withdrew into a world of my own making.

The frustration that followed failed treatments led to anger. My father blamed my mother. My mother blamed the doctor. She insisted that the severe progression of my illness and even the reason I became sick in the first place was because the doctor prioritized the Arab Kuwaiti citizens and refused to sign off on additional tests until it became too late. And I blamed myself: I must have done something naughty to deserve this. The more I withdrew into my world, the more my family rallied around me, speaking louder and writing things for me to read, or clapping to get my attention. They asked, repeatedly, perhaps frustratedly, how much of my hearing was lost. They received only silence in return, a stubborn refusal to answer.

My mother sent my files to a friend in England for a second opinion. A few months after being discharged from the hospital, I still couldn't stand up or walk, so an English doctor prescribed medication for balance to help me regain my mobility. A trip to the Sick Children's Hospital in London followed. There, I was given a large orange hearing aid with headphones awkwardly strapped in a harness on my frail chest. Their promise of amplification did little to improve my world, but they were better than nothing, better than the exclusion of silence. The sounds, however, were difficult to understand, especially garbled speech; and the hearing aid itself was uncomfortable to wear. Searches for cures continued for two more years. Eventually we immigrated to Canada, where eventually I was fitted with my first analog behind-the-ear hearing aids. They were large on my ears, but with them, my auditory world changed once again: I could discern between different sounds, hear my parents' voices clearly, and even—to some degree before closed captioning—understand television.

My family's response to my illness and subsequent deafness is a common one among hearing families. It is a tragedy, a misfortune that plagues the household, when a child is deprived of one of her senses—it is all too easy to see her as "broken." It becomes difficult to face the new reality of special treatment and special education. It was a reality even my younger siblings had trouble grasping. One of my sisters, three years younger than me, would constantly tap on my shoulder to get my attention, never fully understanding why I could not play with her as I used to. Playtime with her and my cousins became a new dynamic: to others, my words were unintelligible, my tears lacked explanation, and there was always a wall between us that I could never penetrate. It was easier to sit on the sidelines and watch.

It was not until we settled in Toronto and I was enrolled in a special class that I realized there were other children like me. Slowly I came to understand that I was hard of hearing and the implications this had for my life. Speech therapy (lipreading) was required to transform garbled strands of speech into articulated words. I was deaf, hearing aids helped me to hear, but speech-reading let me "fill in the blanks." I never formally learned sign language, though I did know its alphabet and used it as a secret code to talk with my friends.

I became obsessed with hearing, mostly out of fear that I would miss out on what everyone else was experiencing. From my perspective, everyone else heard better—even my classmates—and I was constantly struggling to catch up, to participate in jokes or understand song lyrics that others were singing. I used to blindfold myself to train my ears to hear only through my hearing aids. By the time my friends and I reached high school, we had carved our own paths to better adjust ourselves to the world. I was placed in a gifted education program and fully integrated into a hearing school (a process called "mainstreaming"). Some of my friends stopped speaking and wearing hearing aids, preferring sign language as their primary mode of communication, becoming full-fledged members of the Deaf community. A few got fitted for cochlear implants. Eventually I became an outsider in the same community in which I had once found a sense of belonging, because I had become too "hearing" to be "deaf." I had an audiological impairment, yes, but I was not culturally Deaf: I did not sign, belong to any Deaf organization, or participate in any Deaf events. But my "accent," the result of a speech impediment, gave away the fact that I was not truly "hearing." For years, I was stuck in the limbo between two worlds: never quite deaf enough or quite hearing enough to perfectly fit in one side or another.

An Oppression of Difference

When I first began writing this book, the story I had in mind was about how the story of medical and technological treatments for hearing loss displayed a repetitive cyclical pattern throughout the past hundred years. I focused on two crucial questions: What is it about deafness that leaves it vulnerable to the clutches of fakers and frauds, or to medical practitioners who insist on invasive treatment without guarantee of a cure? Why this obsession with "fixing" deafness?

Investigating the evolution of certain kinds of "deafness cures," from promising treatments to obscure fads, I imagined what it would be like if such treatments were imposed upon me, or if I had pursued them myself. As a historian, I am trained to distance myself from past actors, to remain objective and view sources through methodological rigor. The past, however, is an alluring siren, whose emotions remain entrenched in memory. It offers nostalgia and admonition, is marvelous and painful.

The memories of deaf persons leap from dusty pages. Their stories resonate with optimistic dreams of surpassing auditory limits and securing normalcy as citizens. The aching of hearing parents longing to connect to their deaf child. The suffering of children in residential schools subjected to experimentation. The increasing isolation of elderly people as silence threatens to make their once-flourishing lives irrelevant. Fears of exposure, of wires and boxes, technological parasites clutching on one's body. Of stuttering and frustration, lost loves and despair, reverence and redemption.

On a quest to balance the overarching history of medical advancement with stories of deafened people—those who are hard of hearing and not completely deaf—, I decided to return to the archives. My partner and I planned a research trip across the northeastern United States, chasing down leads so I could examine as many archival collections as possible. We borrowed my sister's old hatchback, stuffed all our belongings in the trunk, connected our phones and laptops, and headed out on the road. We didn't know how long we intended to be traveling, but we did have the freedom to unearth new paths for me to explore. As I followed the trails of deaf people, I discovered stories I never imagined I would find. Stories of relationships and hope, as past actors negotiated with medical experts, hearing aid salesmen, and charlatans. Stories about the pressures of normality and what "hearing happiness" meant to those who could hardly hear.

I found myself placed within this history and realized that the decisions I make, and that others make, about my hearing loss are a product of history. This positioning also influences the way I write history. South African poet David Wright, who lost his hearing from scarlet fever at age seven, wrote: "About deafness, I know everything and nothing."[3] I too know everything about deafness from my experiences learning to adjust myself to a hearing world. These experiences allow me to connect with and sympathize with people. But I also know nothing, because the myriad experiences of deafness and hearing impairment are built on a foundation of many layers and threads of historical analysis. What I do know for sure—what I see clearly as a historian—is that deafness is usually a negotiation about normalcy, rooted somewhere between hearing and speech. In my case, this negotiation involved a series of attempts to "cure" my deafness so I could hear better, speak better, *be* better. Masking my speech impediment to protect me from full exposure as a deaf person. Being subjected to an abundance of folk remedies and religious rituals to invoke divine intervention. Adopting various hearing aids designed to amplify my auditory world. Seeing my family physician to inquire about the latest medical or surgical breakthrough I had seen on the news, only to be disappointed upon being told my impairment was too severe for such procedures. Refusing to wear the FM listening system (a frequency modulator that connects a transmitter worn by the teacher to the hearing aid) in the classroom because it denied me the opportunity to train my hearing within a cacophonous world. Understanding what it meant—and means—to be classified with "profound-to-severe" hearing loss.

The space I study and the space I live in are chimerically linked, as they are for other deaf scholars like Brenda Jo Brueggemann and Kristen C. Harmon.[4] Placing my own voice in the narrative serves more than a rhetorical purpose: it is a reminder that despite immense medical and technological advancements over the past decades, deafness continues to be perceived as a "problem" in dire need of a solution. This problem is rooted within what scholar Christopher Krentz defines as the "hearing line": an invisible boundary separating deaf and hearing people.[5] The line creates a conceptual binary to frame identity (of others, or of self) and to negotiate the implications of that identity. An identity whose narrative toggles between deaf and hearing, rooted in a mediated space between disabled and dis/abled, between abnormal and different.[6] It is a problem of the body, of the misunderstood and silent one, a body poked and prodded, inspected and rejected, as it grapples with difference and diagnosis in its pursuit of normality—in spirit, if not in actuality. It is also a political body moving between and beyond borders, deliberately crafting identity within frequently changing tides of time and place. This is a sense of difference, a sense of responsibility and participation that becomes problematized within the demands and constraints of medical intervention.

"Deaf people's bodies have been labelled, segregated and controlled for most of their history," Carol Padden and Tom Humphries declare in their 2009 book, *Inside Deaf Culture*. "This legacy is still very much present in the specter of future 'advances' in cochlear implants and genetic engineering."[7] Within this legacy resides a history of negotiations over unstable identities as much as over informed consumerism regarding health care, the interplay of professional interests, and the expansive role of advocacy. By anchoring myself in this historicity, I outline how the model of normalcy that constructs deafness as an impairment reinforces notions of able-bodiedness and audism—it suppresses hearing variety, stigmatizes faces of inferiority (quite literally, regarding the so-called "deaf face" of mishearing), and demands correction. By occasionally juxtaposing my experiences of hearing loss with past actors discussed in these chapters, I offer a way of thinking about history, particularly how we think and talk about deafness and hearing loss within the constraints of medical intervention, including how to frame deafness—and disability more broadly—as an oppression of difference rather than an impairment.

Normal Conformity

The history of hearing loss is more than a history of medical and technological intervention. It is a history that incorporates ideals of citizenship and philosophy to debate the meaning of humanity and what we consider "normal."[8] "Normal" first appeared, according to the *Oxford English Dictionary*, in 1589 as an implied meaning for doing something in "a regular manner," before becoming widely used in a mathematical context—for the geometric perpendicular line. By the mid-nineteenth century, "normal" was adopted to explain conformity to a type and used as reference points for understanding the context of health and biological functionality.[9] By the 1880s, the term was used with increasing frequency among medical specialists who sought to identify categorical differences between individual measurements or characteristics. "Normal"—and the more general category of "normality"—however, would not enter the popular lexicon until the 1920s and 1930s, when statistical data was translated for public health discourses and popular culture.[10] The Harvard Grant Study of Normal Young Men and the exhibition of the statistically "average" wax bodies of Norman and Norma at the Cleveland Public Health Museum further generated the concept of the normal within the public imagination. As historian Anna G. Creadick argues, normality then emerged as a post-traumatic response to postwar anxieties, serving as a metonym for conformity.[11] Normal thus became the definition of the ideal.

As a way of structuring and ordering the world, normality also became a way to measure and define American bodies and American life, operating not as a binary, but as a way of knowing that often proved impossible to achieve. Yet normalcy was more than a politi-

FIGURE 0.1. Hearing aid advertisements often referred to the importance of "normal living," signifying the need for deaf persons to assimilate into hearing culture. Acousticon hearing aid advertisement, c. 1940s. Kenneth Berger Museum and Archives.

cal or medical criterion, for it represented a comforting cultural standard for many Americans to shape their ideals.[12] Within the space of bodies, especially disabled bodies, normal became "a space for self-cultivation and self-improvement," a nexus for identifying "good citizenship" in the form of paid work and productivity independent of state welfare.[13] It also relied on meanings that intersected with the cultural factors of family, workplace, and public policy, shaping the complexities of hearing loss through normative assumptions of hearing. If good hearing is valued as the ideal, then it drives the assumption that poor hearing—or no hearing—is unacceptable, or at least tragic.

Deafness then, is a cultural construction as much as a physical phenomenon.[14] This is a history of how the Deaf community was co-constructed as a minority linguistic group and *also* a history of how medical treatments of deafness work to enforce normalcy. As historian Douglas Baynton argues, this construction is an expression of broader cultural values about normalcy, including how expectations of medical intervention tend to change in response to expectations of ability.[15] That is, if medicine can provide a "cure" that will transform nonnormal bodies into normal ones, then we are *obligated* to accept it. Indeed, in *Damned for Their Difference*, scholars Jan Branson and Don Miller add that although deafness has been historically categorized as "a symptom to be treated, ameliorated and denied, though never quite cured," this did not prevent ear specialists or patent medicine vendors from offering miraculous cures.[16] This search and attempt for a cure becomes part of a process of delivering a standard of normality, one that made use of medicines, surgeries, speech therapies, and acoustic aids to push forward the ability of the deafened to overcome their limitations. Accept a cure; become "normal." That was the promise, even if all experts could promise was to *amplify* hearing, not necessarily *restore* it.

Even before medicine attempted to normalize hearing loss through intervention, deaf people were subjected to philosophical and linguistic attempts to "unlock their rationality." Seventeenth-century philosophers relied on deaf people to understand humanity: if those "without a voice" were unable to reason, were they even capable of exercising free will or believing in God?[17] And if they were taught an alternative system of communication, could we uncover divine systems of universal language that connected all of mankind?[18] Though the first formal systems of sign language were developed in Spain and France during the 1790s, they were aimed at offering deaf persons communication to open their mind, to be receptive to reason and the word of God. Scores of residential schools and church programs followed in the nineteenth century, marking the beginning of a community of deaf persons and the possibility that the great wall between the hearing and the deaf could be broken down. The uniformity of sign language that eventually developed in these schools became an essential step in building a culture that continued long after graduation: organizations, newspapers, events, churches, and employment within the community formed the basis of what would come to be a (capital-*D*) Deaf culture.[19]

Not everyone perceived Deaf culture as an enriching, exciting, and unique part of American culture. Within the tenets of late nineteenth-century scientific naturalism, sign language occupied a low rung on the scale of evolutionary progress, the signs mere gestures of bestiality, with the deaf-mute isolated in silence and lacking rational thinking. It was the most "savage" of languages, as Baynton emphasizes, perceived as a supposed link between the animal and human—speech was the "natural" way of civilized humans.[20] Such ideologies seeking to Americanize marginalized groups by assimilating them to mainstream society and the democratic process were rooted in anxieties over the increasing number of "inferior races" that, without checks, would proliferate undesirable defects, diseases, and deviances to weaken the constitution of the nation. The tide of anti-immigration and eugenic sentiment shaped legislation and enforcement, while defining disability by reflecting cultural anxieties over what it meant to be a citizen.[21] Marriage between Deaf persons was a threat to the hereditary makeup of the nation. The Deaf were expected to give up their separate community, their "clannishness," and absorb values that reflected the vision of "normal" America. That is, they were to be godly, educated, civic-minded, and hearing.[22]

For deaf persons, being an ideal citizen required assimilation into the hearing world, rather than securing belonging within the confinements of the Deaf community and becoming "foreigners in their own land." Campaigns to assimilate and acculturate deaf people were spearheaded by the ideology of *oralism*: a framework advocating the perception that speech and articulation training were modes of instruction superior to sign language.[23] As disability historian Lennard J. Davis elaborates, the basic argument of oralists was that "if only deaf citizens could speak and understand English, there would be no problem for them or the larger community. Thus, deaf people are schooled arduously in lip-reading, speech therapy, and the activities associated with the oral/aural form of communication."[24] Once oral educators allied with politicians and medical specialists, they became a network of collective action making colonialist decisions to gain and exert control over deaf bodies.[25] Over the course of the twentieth century, oralism became the doctrine of a powerful biomedical network that completely shifted the meaning of deafness to a medical condition that required permanent restoration of hearing.[26] Tying speech to normality, oralists offered their own "cure," an alternative, if not an addendum, to medical solutions and hearing aids. Those who failed to successfully speak were not only "oral failures," but incurables who had failed to pass.

Oralism is an important aspect of the history of deafness cures. Under the assumption that deafness is a medical problem in dire need of solution, professionals crafted deafness as an impairment that could be targeted on various fronts: medical professionals offered prevention and treatments, surgeons presented invasive manipulation, scientists and engineers developed advanced technologies, and oral educators eliminated social and educational barriers. Their goal was to eliminate the handicap of deafness, and each group, as Susan Burch asserts, "sought to *normalize* Deaf people according to mainstream values. Enabling Deaf people to talk, and ideally, hear better, would supposedly 'restore' them to the broader world."[27] If there was no medical cure, physicians promoted oral education and acoustic devices as the last clinical approach for treating deafness, a process traced to nineteenth-century aurists (ear specialists) who promoted a combination of medical treatment and social welfare. Continuing into the twentieth century, this medico-educational arrangement expanded to include lip-reading classes and hearing aid assessments as tools for the deaf to portray their normalcy.

Presented as restoration—of self, of bodies, of hearing—oralism, as Baynton argues, "was adopted not because it accomplished more but because it seemed to better answer the anxieties and concerns of the time."[28] Oralism was the normal condition, used to establish "not just the existing, but the desirable and the right."[29] Educators defended this stance, arguing, as John D. Wright did in 1924, that supplying a "normal environment" and advocating normality was an effort to make "deaf children wholly normal," even if such efforts could not benefit all their pupils.[30] "I have always believed," wrote educator Hilda Tillinghast, "that in as far [sic] as it was possible we should make our deaf children more like hearing children."[31] By holding their deaf pupils to cultural standards of normality and attempting to situate them in a normal environment, educators incorporated broader notions about good citizenship, to overcome the isolation conveyed by their deafness and become productive citizens. They merely provided a structured environment for integration. Adjustment was the child's responsibility.

Of course, conformity to the normal required being a good citizen who was productive and self-sufficient. In the 1910s, social organizations such as the New York League for the Hard of Hearing lobbied for new categories to distinguish those who were hearing impaired from illness or injury and relied on hearing aids and lip reading—the "hard of hearing," or "deafened"—from those who were d/Deaf without speech. Far from concerning themselves with progressivism, the league constructed their identity as "hard of hearing" in their quest to create a medical discourse surrounding their condition.[32] Raised in a hearing society, they experienced the loss of a sense, and thus considered themselves to be "normal" citizens, but with a medical condition (in some cases, temporary) that handicapped the full range of their abilities. As described in the league's annual newsletter, members were not "deaf and done for," but rather "hard of hearing and hopeful."[33]

Expected to pass as hearing and conform to social expectations to assert their normality, the deaf relied on acoustic aids, medical therapies, speech therapies, and a host of unconventional therapies that promised grand miracles but failed to deliver. Advertisers honed the message: by establishing conformity through normality, the problem of deafness was a problem of better living. To fail to assimilate into the hearing world was to be "un-American."[34] To fail to conform was to give up one's quest for hearing happiness. Images of deaf people relieved by fixing their defect or enhancing it with hearing aids; the transformation from sullen and moody to content and cheerful; joy and laughter not tears and despair: these were the messages continuously sent to the deaf selling products and promises of hope and cure. Some were genuine, others outright lies. Of course, seeing these messages, parents of deaf children and deaf individuals alike want to fix the defect.

Deafness is revealed only when engaged in conversation with a deaf person or in a group of signers. Yet deafness has no meaning other than what is assigned to it, meanings created by hearing people projecting their own ideals of hearing and normality, or even meanings ascribed by deaf persons themselves. The polarized narratives of deafness and hearing often closely mirror American history. It is important for us to chart the battlegrounds, however. Not only to understand the proliferation of frauds and fads in deafness cures, but also to grasp the transformative ways that hearing was historically imposed upon the deaf and deafened, to gain insight into the ways these individuals negotiated how and *whether* they should "live with" their deafness or opt for a "cure." In so doing, we can trace the history of how deafness came to be medicalized, and how the pressures to conform to the medical norm of ability led deaf individuals, and especially parents of deaf children, to adhere to the expertise of medical practitioners and explore treatments with faint possibilities of success.

Deaf Sentence

In her memoir, Donna McDonald recounts her realization of "an apparently unbated prejudice" surrounding deafness that led parents to perceive news of their child's diagnosis as a death sentence. Why perceive the situation as "the death of hope and prospects for their child, when the facts show that far from being a death sentence, the diagnosis of deafness propels a child into a different life, not a lesser life?"[35] I read this passage and thought about the story my mother told me about my name.

When a child is born into a Sikh family, the ritual of *naam karan* is organized as soon as mother and child are healthy enough to attend the gurdwara. This is the naming ceremony, a formal presentation of the newborn to the holy presence of the *Guru Granth Sahib*. Upon reciting hymns, the holy *hukam* (command) is delivered: a name starting with the first letter of the hymn of the hukam is proposed as directed by divinity. My letter came up as "S." Disagreeing over what to name me, my parents decided to choose another letter, give me another name, another identity—perhaps in defiance, I don't know. Then a few years later, I became ill, and unspoken words hung in the air: was this illness a punishment? A death sentence of sorts, where the "real" me had died and been replaced with a changeling, a false identity masked in silence? This silence of hearing loss, McDonald writes, eventually paved the way for a different sort of silence, the "silence of lost stories, invisible stories, unspoken stories." To this, I add the silence of possible identities and different selves.

On the subject of the formation of self-identity, Daniel J. Wilson emphasizes that an "important step in passing in the world of normal was first to 'pass' in one's own thinking."[36] Passing is the process whereby people conceal social markers of impairment to avoid cultural stigma. It's the wearing of masks, adopting, even absorbing another identity, pretending to be complete. By (carefully) managing self-presentation, passing allows individuals with disabilities to portray behavior and social characteristics perceived as "normal"; passing then, "expresses, reifies, and helps create concepts of normality," even if those concepts are hardly definite.[37] I could imagine myself as hearing, but easily be betrayed by the "whistling" sound of my hearing aid. Never mind if those whose hearing was offended by the piercing blast were sympathetic; the embarrassment over the reveal was so deeply rooted in my psyche that I used to deny the exposure, pretending that I was also offended by the awful sound. I wasn't, of course. I couldn't even hear it. But in my mind, I was offended by my *self*, angry that I had slipped and dared to expose my deafness in the world. I got over this in adulthood, acknowledging and apologizing when the whistling occurred and promising to get it under control.

Speech was another story. The embarrassment of mispronouncing a word, seeing a word such as "reciprocal," or "hypotenuse," or "Copernicus," and having no idea how it sounds. To try, only to be met with laughter. It's an emotional spiral. Every once in a while, I catch myself repeating the word silently to myself as a mantra, even though it's been years since they first laughed at my mispronunciation. Have I gotten over it, as I have adjusted to my deafness? Even though I tell my class on the first day of lectures I am deaf and wear hearing aids, and that they should not shy away from correcting any mispronunciations on my part, a flush of redness envelops me when they do. They apologize; I'm grateful for the lesson, but then walk away feeling incredibly stupid. It won't matter how much I pretend, how many times I write down in my lecture notes in red ink giving the correct way to pronounce a word—the moment I am exposed, I am raw and vulnerable. I may have told others about my mask, but it has slipped. And I cannot hide that I am not "hearing" but simply pretending to be so. They have caught me. I am an impostor.

In her 1949 autobiography *Hearing Is Believing*, Marie Hays Heiner writes: "To understand (hear) and to understand (comprehend) appears as the one and same thing to most normal hearers. By this yardstick, the hard-of-hearing often appear as stupid."[38] In self-defense, we avoid this categorization by wrapping ourselves in vanity. Heiner elaborates:

> To some extent, nearly all we deafened are vain. Perhaps it's because we're sensitive. Deafness threatens our vanity more than a blind eye or a lacking limb because people in general confuse hearing, a physical function, with understanding, a mental process. By this reasoning a person who hears poorly is held to understand or think poorly. . . . Because a low hearing pick-up is still confused with mental slowness by the hurrying, hearing world, we're frequently dismissed as dumb when we are only deaf.[39]

Sociologist Erving Goffman addresses this issue of self-stigma, explaining how stigmatization and shame drive a person to "correct what he sees as the objective basis of his failing."[40] By seeking out repairs and correctives for their defect or deformity and being willing to adopt extremes to correct their "blemishes," the stigmatized person tends to fall victim to fraudulent services and remedies. They become more prone to victimization if their quest for a cure is driven by self-stigma or cloaked in secrecy. And the extreme measures employed to bring the broken, blemished self back to normal are extended even further by a devotion to mastering areas of activity that would disguise any shortcomings—much like how I used to blindfold myself to train my ears, or how I used to write short stories so people would focus on my writing rather than my speech.

I was six years old when I received my first pair of behind-the-ear hearing aids, the beginning of my feelings of self-consciousness and shame. They were a heavy burden pressuring on the sides of my head, making my ears stick out. I was convinced they drew even more unwanted attention to my deafness. My hair, tied in a long braid as per Sikh tradition, did little to disguise them, or limit the snickers, puzzled glances, and finger-pointing from younger children. Goffman contends that the stigmatized person's failing is pointedly exposed through children's stares. This was certainly true for me. "They help me hear," I would reply to children's never-ending questions about the things behind my ears (though I too was merely a child struggling to grasp that I was not "normal"). "So I can hear like you." The second explanation was always a lie. My younger self knew that no matter my explanations, no matter how many times I took out my hearing aids to quiet inquisitive questions, as soon as the finger-pointing commenced, I could not hide my defect.

Clunky though they were, my hearing aids were a great improvement on the first commercially available models. First developed in the 1900s and modeled upon the principle of the telephone, these aids relied on battery-operated carbon transmitters and receivers, and were far more advanced than the mechanical ear trumpets, conversation tubes, and acoustic fans of the past. Yet these early technologies were far from discreet. Large, bulky, and requiring heavy batteries that had to be charged frequently, they were frustrating, if not obnoxious, for most users. Poor sound transmission, inadequate amplification, and general discomfort led many users to discard their products; some, out of sheer embarrassment, refused to wear the instruments in public, fearing it would draw more attention to their deafness. Even when the vacuum tube was introduced for hearing aids in 1920, making it possible to improve amplification and miniaturize the aids, many deafened persons refused to wear their devices.

By the 1950s, hearing aid companies were strategizing how to sell their products to reluctant customers who not only complained about discomfort, but also candidly expressed their embarrassment about exposing their impairment. Indeed, at the June 1950 meeting of the Better Business Bureau in Washington, D.C., Irving Schachtel, the president of Sonotone, a hearing aid industry leader, expressed an obvious truth: "that nobody wants to put on a hearing aid, and that hearing aids are the most difficult to sell."[41]

If deafness is highly stigmatized, perhaps it is because its technology and modes of communication make visible its nature as an otherwise invisible impairment. No wonder aids to hearing have always been designed to be hidden: concealed within the skin or body, or disguised as furniture or ordinary domestic objects. In the nineteenth century, wealthy people had the option of purchasing mechanical aids that could "disappear" when not in use: fans, urns, thrones, headbands, and walking canes.[42] Twentieth-century innovations included specialized clothing—hats, undergarments, and belts—to hide the devices on the body. Women had the additional option of purchasing jewelry, buttons, clips, and hair barrettes to discreetly tuck away their hearing aids.

Twentieth-century advertisers certainly highlighted this need for discreetness, boldly proclaiming that "deafness is misery, good hearing a joy." This perception has not disappeared with the advent of new technologies or even with tolerance for differences. In 2015, for instance, the Australian hearing aid company Victorian Hearing came under fire for deaf-shaming when its advertisements surfaced with the tag-line "HEARING AIDS can be UGLY" next to a photo of a woman wearing a prawn behind her ear.[43] A year later, gossip blogs were buzzing over reports that Caitlyn Jenner was photographed wearing a hearing aid, ridiculing her "transition" to deafness.[44]

Invisibility is a popular selling point for hearing aids. It is better to contour and fit the technology into the intricate parts of the ear, it seems, than to expose the ugliness for the world to see. The combined effects of stigmatization and the growing authority of medical experts meant that deafened persons often downplayed their hearing loss to ease communicative tensions. We've all seen the cartoons mocking the deaf person who repeats "huh? huh?" Or hurled the insult "what are you, deaf?" at those who failed to listen. Invoking laughter at the miscommunication between the hearing and the deaf has been a cliché, but also an effective joke. So much so, that some deafened persons turned to the "plausible clap-trap of the deafness-cure quack" out of desperation.[45] Others feared receiving the "death verdict"—the deadness of the ears—from a physician or otologist, as was the case of Charles J. Gotthart of Chicago in 1925.[46] It meant that there was no hope and "no phone on the market will help," and life would be despairing. If hearing aids were embarrassing—if not completely useless—then it was better, perhaps even easier, to risk one's ears at the hands of an (un)experienced, innovative practitioner, or to purchase a prettily packaged "special" oil, rather than to live with deafness.

This message was so powerful that even as safer surgical procedures and improved technologies were developed to amplify hearing, the audist connotations remained steadfast, as did unscrupulous quack peddlers who repeated their same bold declaration of "HOPE FOR THE DEAF." A 1927 advertisement for Dr. W. O. Coffee advised "25,000 sufferers from deafness" not to neglect their troubles and offered a starling new and free offer for treatment.[47] In 1991, a television commercial for Miracle-Ear's new "Clarifier" aid encouraged "you or someone you know [who] thinks they may be suffering from hearing loss" to call the toll-free number to receive helpful information and a booklet on "better hearing."[48] Just as the Miracle-Ear was so "tiny, it's practicable unnoticeable," so too were the artificial eardrums of the 1920s, small acoustic aids inserted into the ear canal to resonate sounds to the eardrum.[49] Referred to as "telephones for the ear," "sound discs," "earphones," or even "tiny megaphones," at their heyday artificial eardrums were branded as one of the most innovative medical developments of the time. They promised invisibility, assuring their customers that deafness could be discreetly hidden and that "deaf persons need no longer despair."

"Why be deaf?" advertisements chimed incessantly.

Frauds and Facts

In 1886, the *New York Times* editorialized: "Deafness is the commonest of the more serious physical infirmities to which flesh is heir. In this country it is particularly prevalent."[50] Nearly fifty years later, a physician from Long Beach, California addressed the worrisome statistics of deafness: one study discovered nearly thirty-five thousand Americans were deaf. Another study found out that of a million people who received a hearing test, 6 percent had significant hearing impairment. Yet another study reported that three million Americans suffered from some degree of hearing impairment. Deafness then, was a serious problem that required invasive state intervention. After all, as the physician stressed, "the three great public health problems confronting the world are heart disease, cancer, and deafness."[51] Today, there are approximately thirty million Americans who have some degree of hearing impairment.[52]

This notion of deafness as a worrisome public health problem concerned families and physicians alike. Despite remarkable advances made by Renaissance anatomists outlining the structure of the ear and its parts, physicians generally agreed that the ear was too delicate an organ to be operated on. By the 1860s, when the specialty of aural surgery—an archaic term for otology, later otolaryngology (ear-nose-throat specialty)—was beginning to become more refined, surgeons were able to apply greater clinical acumen and pathological anatomy to understand and treat ear diseases. Even as prominent British and American otologists gathered in 1876 at the first International Congress of Otology in New York, the medical advancements they made were insufficient to explain how or why hearing loss occurred in cases where there were no structural or physiological defects.[53] Diagnosing deafness was a challenge, curing it insurmountable.

This did not, of course, stop people from seeking out treatments to improve their hearing power, whether for themselves, families, friends, or neighbors. They borrowed ideas from medical books imported from Europe, tested ancient recipes copied from health manuals or druggist handbooks, experimented with popular electrical machines capable of transcending the "deadness" of ears, and even haggled with sellers over prices of nostrums. Others turned to faithful family remedies: using decongestants such as tobacco smoke to clear out the head, filling ears with fluids, wearing tight caps, plugging ears with cotton, or trying out various purgatives and laxatives. In her description of families with deaf children in the antebellum South, historian Hannah Joyner describes how the prevalence of deafness as a tragic misfortune affected the medical choices hearing parents faced in their attempts to "fix" their child.[54] Herbal remedies were frequently advised by family members, as were more invasive suggestions, such as leeches and blistering behind the ears to draw out toxins. Furthermore, Joyner explains, since "physicians felt that the horrors of deafness were worse than any medical treatment, they were willing to operate experimentally on Deaf children" and test out theories, often without parental permission.[55] Dragged to specialist after specialist and forced to undergo dangerous, uncertain procedures, deaf children (and adults) ended up feeling more alienated from their families.

Patent medicines and restorative tonics were alternative, inexpensive options for treatment. The parlance of "patent medicine" did not necessarily mean a product was patented. Rather, the term originates from seventeenth-century letters patent granted by the English Crown to legitimatize an inventor's monopoly over a secret medicine formula. It then became corrupted and exploited after trade disruption during the American Revolution, as manufacturers misused the term to protect their proprietary markup. These medicines often promised full restoration of hearing, their makers boasting in newspaper advertisements, mail-order catalogs, and flyers about the miraculous features of the product. They were especially prominent during the early twentieth century, when distrust and frustration over traditional medical therapeutics drove people to seek alternative treatments.

Deafness treatments have left many traces in the historical records: families or individuals attempting to treat their hearing loss through different forms of medical, religious, or technological methods; case studies of former pupils at residential schools for the deaf whose hearing was restored in adulthood after invasive surgical intervention; home recipe books with a remedy for deafness using black pepper, carefully handwritten with the note "it will give immediate relief"; ear ointments made with organic ingredients such as skin of serpent boiled in wine, fat of fox's lungs, egg yolks, or goose grease; medical textbooks listing descriptions of experiments using burning caustics, blistering, setons, or hammering of deaf children's skulls; excitement, uncertainty, and skepticism lingering over sensationalized treatments promoted in newspapers, especially if experimented on children—hypnotism, audiphones, light rays, vibrating tubes, and other apparatuses all had their moment in the spotlight. Letters from working-class deaf individuals across America narrated their experiences while hopefully inquiring whether a new artificial eardrum, ear trumpet, hearing aid, or surgical procedure was worth splurging their hard-earned wages on. There were even appeals to famous inventors like Thomas Edison and Alexander Graham Bell to commercialize problem-solving inventions for one's hearing defect.

These stories narrate deafness as a myriad of experiences: late-deafened, progressive deafness, deaf from birth, "deaf and dumb," orally deaf, hard of hearing, partially hearing, partially deaf, Deaf, prelingually deaf, conductive deafness, "nervous deafness," adventitious deafness, hearing impaired, and deaf. With each experience came an approach to dealing with it, and a series of negotiations between managing the hearing loss and attempting to treat it. One attempted either to restore lost hearing, or to shock (sometimes literally) "dead" ears to create, or make use of, residual hearing, even in ears that had never heard sounds. Of course, despite the boastful claims of advertisers, none of these cures were certain in their claims to alleviate deafness. Nor were they painless. Yet that did not stop people from either trying or recommending these treatments. If there was even a glimmer of hope that hearing could be fully restored, then the cure was worth it. And if a treatment was heralded as "new hope for the deaf," promising permanent hearing amplification, it was popularized, exploited, and sensationalized as a catch-all cure for all kinds of deafness—despite the protestations of ear specialists insisting that such treatments only benefited a small percentage of cases.

This is what it's like to be deaf in America: any cure is better than no cure.

Beginning in the nineteenth century, deafness cures were legion. The field of aural surgery was, like many other medical specialties, on the cusp of professionalization, battling internal discord and occupational criticism as its practitioners pushed for advances in diagnosis and treatment while counteracting public opinion that deafness was medically incurable. The aurists who professed a cure were regularly dismissed as pretenders, quacks whose fallacious promises were mere cloaks for purse milking, even if they were medically trained and properly qualified. Moreover, the difficulty of adequately assessing the delicate and intricate parts of the ear for diagnosis made treating hearing disorders more challenging. As one physician commented, if all aurists were honorable men, the varied nomenclature and contradictory diagnoses, to say nothing of the bickering and bashing in the profession, only served to ripen the social distrust of aural surgery.[56]

One problem with this distrust of aurists is that it paved an inviting path for medical peddlers and unscrupulous companies to promote "total cures" for deafness, offering hope where physicians failed. Thousands of testimonials supported these peddlers' practices and skills, even though some (if not all) of these testimonials were fabricated, recycled content. They were superb marketers and brilliant entrepreneurs who knew how to manipulate for profits. As Eric Boyle outlines in his study of quackery, advocates "of questionable medicines marketed their wares to consumers by appealing to vanity, instilling fear, creating hope, and promoting the freedom of unfortunate victims of diseases."[57] They promoted their products with attractive advertisements using bold headlines such as "HEAR-ING RESTORED!" While many of the benefits of these advertised products were certainly exaggerated, coupled with sales tactics that included free gifts and trial loans, they made deafness cures available easily and discreetly. Bottles of tonics or a box of artificial eardrums could be cheaper than a visit to an aurist; with an added bonus, they came with guarantees of relief or your money back.

"A thousand-to-one chance that the wonderful gadget may be a tenth as good as it as it claimed to be seems to be worth a dollar to many people," wrote American physiologist Hallowell Davis in 1954.[58] "But actually there isn't much of a chance," he continued, as the claims of most gadgets to restore hearing are nothing more than "pseudo-scientific smokescreen." Fakes, follies, and frauds frolicked in the market, their popularity surging with each new fad. This is not to say, of course, that *all* deafness cures were meant to dupe the public.

Nor were cures confined to paramedical realms. Indeed, even aural specialists innovated with treatments that challenged the status quo of their profession, though most of them advocated strict regulation to prevent the treatment from becoming another mar-keted "cure-all." These calls for regulation are historical clues as to how blurred the lines were between reputable and quack treat-ments. British surgeon Joseph Toynbee's artificial eardrum design, for instance, was initially marketed only to medical professionals. Borrowing the design, American manufacturers marketed their own versions directly to consumers, bypassing all medical authority by relying on consumers to diagnose their own deafness. These latter versions were crudely constructed, ineffective, and even dangerous, leading Arthur J. Cramp, the first director of the American Medical Association's (AMA) Propaganda Department, to declare that the "number of quacks and faddists who defraud and deceive the deaf-ened is large, considering the restricted field in which they work.

FIGURE 0.2. "Hearing Styles Through the Years," Sonotone advertisement. Central Institute for the Deaf Collection, Series 3, Bernard Becker Medical Library, Washington University School of Medicine.

Most of the practitioners in this line are crude charlatans; a few possibly come from that 'twilight' zone of medical practice, where it is difficult to differentiate between the quack with a scheme and the visionary with a theory."[59]

This need to maintain professional jurisdiction by labeling competitors as charlatans and quacks builds on a classic medical impulse to define legitimate medicine.[60] During the twentieth century, otologists defended their jurisdiction over deaf bodies by dismissing all unconventional therapies proven to be ineffective, labeling them as quackery in their attempts to define standards and expectations for deafness treatments. Their goal was to reduce false hopes for patients while paradoxically also offering greater opportunities to eradicate deafness with groundbreaking surgical procedures. They distanced themselves from quacks by asserting their gentlemanly attributes, insisting that their scientific authority enabled them to not only treat, but also *cure* deafness. Yet any grand claims of cure also incited criticism from their professional brethren, along with accusations of quackery, especially if they arose from inappropriate or unorthodox means. So, while we may place Curtis H. Muncie's "finger surgery" and Julius Lempert's "window operation" on two different calibers of surgical effectiveness, both men centered their flamboyant personalities to promote their innovations as widely as possible, disregarding any appeals to their gentlemanly status. Breaking with expected decorum did little to damage Muncie's and Lempert's reputations as deafness curers, for hundreds of eager patients still went to their offices seeking treatment.

Much of this preoccupation with a deafness cure was due to the variety of therapeutics available to the patient-consumer. By the twentieth century, hearing-impaired persons had plenty of choices for self-treatment, which could be advantageous, for they could select from an abundance of options, but was also troublesome, for it made it difficult to determine a treatment's efficacy. Clara B. Seaman of Ithaca, New York, for instance, wrote to the AMA in 1919 requesting advice to help her select treatment: "There are so many earphones, good, bad, and indifferent and so many kinds of drums, etc., that one who is deaf spends a lot of money, time, and nerves trying them out."[61] She debated asking a physician for advice, but acknowledged that "it is so hard to find a physician . . . who knows enough about these devices to know whether they fit your case or not." Even if a physician or otologist was able to offer advice, however, they often faced frustrated and discontented patients who wanted nothing less than a full restoration of their hearing. As otologist Max A. Goldstein describes from his own experiences, the "deafened adult is easily discouraged with his handicap, often fails to take the advice of the otologist, is frequently subjected to unnecessary and futile treatments, spends his hard earned savings in going from one medical office to another, and finally, hopelessly, turns to the 'deafness cure' in the daily press or other advertising medium as a last resort."[62]

Each time the media reported on a new deafness cure, a new surgical technique, or a new technological solution for hearing loss, there came the expectation that deafened people were required to seek out these cures. This was especially prevalent after the development of audiometry in the 1920s, which created the "normal curve" for hearing and outlined a new set of physiological categories for deafness.[63] Sociologist Laura Mauldin explains that the triumph of medicine and technology often created pressures in patients to conform to treatment, but it also led patients to resist treatment choices that they felt were forced upon them.[64] This tension, which Mauldin terms "ambivalent medicalization," is a struggle over conflicting expectations of treatment that allows both good and bad consequences to exist: it is a discourse about fixing broken bodies, control over disease process, and also a constant negotiation between hearing and deafness. At the heart of these struggles is the question of what a cure entails: about who holds authority to cure deafness, who has a right to offer a cure, and who decides whether the cure is effective.

For Deaf scholars, the very concept of medicalization is problematic. Under this model, deafness is constructed as a personal tragedy, a disability that requires the deafened to adopt all and any measures to pass. This runs contrary to the experiences of many d/Deaf people, who perceive their deafness as part of a healthy community that uses sign language as a form of communication and conveys a positive outlook on being deaf. Deafness, they insist, is not a disease or disability, thus any (forced) medical interventions to fix, correct, or cure is inappropriate.[65] Yet, as scholars Susan Burch and Kim Nielsen have argued, Deaf community leaders themselves have historically encouraged Deaf persons to assimilate into hearing society to appear as normal as possible in order to ward off associations with "disability" or "deviance."[66] The National Association of the Deaf, a civil rights organization of and for d/Deaf individuals, even sought to maintain marriage autonomy for its constituents, defending them against eugenics discourse by encouraging those with hereditary deafness to voluntarily abstain from marriage.[67] Since deafness could be misinterpreted as mental retardation, insanity, or criminality, passing as "normal" allowed deaf people to protect themselves and avoid the stigma of disability.[68]

As this book reveals, this need for self-protection led thousands of deaf Americans to seek out treatment of one kind or another. Some individuals found themselves swayed by the promise of a quick deafness cure, while others had treatment forced upon them. Perhaps all hoped—even in the slightest—that the treatment would restore their hearing, whether at the advice of an otologist, or at the behest of a salesman. Even if historically—and actually, even now—the demarcation between quack cures and legitimate cures is difficult to pinpoint, the demarcation itself is a product of history: a history of lived realities of deafness, and more broadly of disabilities, and the expectations as well as the limits of medical intervention.

The Follies of Fads

At the height of their commercial success, deafness cures were discussed in periodicals, their merits debated among medical practitioners, educators, laymen, and families with deaf children. Portrayed as the latest in modern scientific breakthroughs, these therapies signified that deafness was not monopolized by any one tradition at a time or dominated by medical practitioners, any more than medicinal "cancer cures" offering therapeutic pluralism were shaped by class and geographic location.[69]

The "name of the deafness-cure quack is legion," declared Arthur J. Cramp. "Some," he continued, "carry an alleged cure for deafness as a 'side-line' for other medical fakes they may be exploiting; some sell on the mail-order plan a worthless 'course of treatment,' while still others, and those in the majority, dispose of, at exorbitant prices, devices that are trivial, worthless, and often dangerous."[70] As Boyle argues, Cramp's campaigns of surveillance attempted to combat medical quackery at all social levels, including regional and national advertisements, and unverified claims made by medical schools, individuals, and organizations.[71] Even hearing aid companies were scrutinized for dabbling in "side-line" cures by marketing "cure-all" electric massagers to consumers whose deafness was inadequately improved with a hearing aid. All products suspected of being fraudulent were reported to the AMA for investigation, with reports printed in the association's journal; the most notorious products, the get-rich-quick schemes, the huckster hokum, the fakers and fraudsters, were later compiled by Cramp in his five-hundred-page volume *Nostrums and Quackery*.

These antiquackery campaigns were also crucial to the passing of legislation to regulate the advertising and sale of health products. The 1906 Pure Food and Drug Act specified that the labels of proprietary medicines should not be false or misleading in terms of identifying ingredients used in the mixture; it had no such requirement regarding claims of efficacy. Stricter federal regulations throughout the 1930s attempted to close legal loopholes exploited by proprietors, including by targeting electrotherapeutics suspected or proven to be fraudulent and the interstate sales of medicine, neither of which was covered by the 1906 act. By the 1950s, the FDA had revived its antiquackery crusade, increasing its educational campaign, working with the Federal Trade Commission and the Better Business Bureau to press Congress for greater regulatory action, especially against misleading advertisements. Between 1934 and 1976, for instance, the FTC issued sixty-six orders against manufacturers of hearing aids for false or misleading claims. Forty percent of those orders were issued in the 1950s, during the height of bureaucratic regulation against medical quackery.[72]

These regulations also targeted the advertising industry. Indeed, as E. F. McDonald Jr., president of Zenith, expressed in a private letter in 1952, nothing "has shaken public confidence in the hearing aid industry so much as the fraudulent and/or misleading advertising and merchandising practices."[73] The problematic issue centered on the pricing and promotion of hearing aids, not necessarily the technology itself (although Zenith was criticized by its competitors for participating in deceptive trade practices by creating an inferior instrument, the $75 Royal Hearing Aid, to steal the consumer base). The desire to sell hearing aids to skeptical consumers was so substantial that even hearing aid dealers were accused of being specifically trained to manipulate their consumers by revealing, if not exaggerating, the dangers of hearing impairment in attempts to close a sale. While some advertisements may have more faithfully reflected the values and preoccupations of advertisers, advertising agents, and copywriters, they nevertheless framed Americans as potential citizens whose lives would be dramatically improved with the latest product.[74] Additionally, though some adverts urged consumers to consult a physician before making a purchase, the implicit message was that *not* wearing a hearing aid was a failure of self.

These commercial processes became capitalist expressions of how to normalize hearing loss, and were mediated by advertisers, salesmen, and medical experts, groups driven by the urgency to "fix" through technological intervention and shift the deaf person from citizen to consumer. They were essential in shaping expectations of health governance by articulating a medical ethos of aural rehabilitation and defining the parameters of medical quackery—and in so doing, they dictated patterns of consumption among deaf people that made them reliant on consumer goods as a way to pass as normal. Even the hearing impaired who shopped for healthy treatments engaged with aspects of critical consumerism, wrestling with rising prices, multiplying choices, misleading advertising claims, and questionable adherence to safety standards.[75] Moreover, market choices for specialist care were further complicated by the fact that otolaryngologists faced competition from a host of other "specialists" who claimed that their innovative techniques produced greater results despite lacking supporting empirical evidence.

Quackery and distrust, then, characterized this commerce in deafness cures, as did hope and desire. Commercial processes were expressions of how to normalize hearing loss. Mediated by medical experts, advertisers, salesmen, and social workers, these cures highlight the fluid boundaries that existed between health care practices and the many ways consumers attempted to regain control over their health. They also emphasize repeatedly the importance of a cure for deafness, the necessity to correct a hearing defect.

Is this push for a deafness cure due to the fact that as a society, we are obsessed with stories of people picking themselves up by their bootstraps and overcoming their limitations to achieve success as a "normal" individual? We are obsessed even though these stories diminish the fact that the framework of normality is relative. This is especially the case for people who "overcome" their disabilities: no matter what kind of success they achieve as *a person with a disability*, the disability is first seen as a marker between the boundary of normality and disability. We marvel at how people pass as normal by hiding the signs of disability.

Recently, I searched YouTube using the phrase "deaf person hearing for the first time." It returned more than 100,000 results. The most popular video is from 2011, a ninety-two-second clip of twenty-nine-year-old Sarah Churman, born deaf, having an Envoy Esteem Implant (the first implantable hearing aid approved by the U.S. Food and Drug Administration) "switched on."[76] When the technician turns on the device and asks Churman if there is any difference in sound, she nods and bursts into emotional tears of joy and laughter. It is an astonishing and authentic reaction, especially when Churman, through sobs, exclaims, "My laughter sounds so loud!" Filmed by Churman's husband, the video immediately went viral and was picked up by several news outlets. Churman made appearances on *Today* and *Ellen*, promoting both the implant and her overcoming narrative. She published a memoir, *Powered On*, describing her transition to the world of sounds and addressing accusations that due to the legibility of her voice, the video was a hoax.[77]

Hoax or no, these videos are misleading. Implantable hearing aids and cochlear implants (which deliver electric signal directly to the auditory nerve) do not restore "normal" hearing; they amplify sounds but do not improve acuity; they facilitate speech-reading and make it easier to distinguish certain sounds. They are not effective for all forms of hearing loss, and even for those whose hearing they improve, the increase in hearing sensitivity and acuity comes gradually rather than all in a dramatic rush.[78] Wearing an implant does not mean a deaf person is *suddenly* able to hear and speak flawlessly—and yet this is the image presented by the media. No wonder then that people are confused when deaf persons refuse to opt for the implant surgery or when hard of hearing persons are not interested in newer technological advances.

Since their first introduction in 1978, cochlear implants (CIs) were propagated as the ultimate cure-all technology for deafness—so much so that Deaf activists warned it would lead to cultural genocide and the eradication of a minority group. Arguments by leading Deaf scholars claimed that the CI was not a reflection of medical progress or technological development, but another instance of the historical oppression of Deaf culture, the ultimate denial of deafness, and the most serious threat to sign language and Deafhood.[79] It did not help matters that the CI emerged just as d/Deaf communities were banding together for the right to a unique linguistic culture and community.[80] While the CI charted yet another battleground, this does not mean that all Deaf persons view medical technologies as a threat against their culture. Some consider cochlear implants as one of many tools that will enable them to navigate through the hearing world.

The "switch-on" is a powerful moment. It's a moment intricately tied up with what historian Jennifer Esmail terms as the "rhetoric of curability," a promise and a reality, of trust in the miracles of modern science.[81] An expectation not completely impossible, that undesirable ailments will eventually disappear, checked at the moment of conception, or else eradicated by the force of technology. The switch-on is a way of thinking about the link, the tension, the obsession, between "normal" and "not-normal" (better than "abnormal"), an ever-present commentary on the tragedy, the sorrow and shame, of difference, on the broken who are, or should be, disposable if incurable. Perhaps it is grotesque to nitpick on the obsessive commentary, perhaps the binary is a false one, manufactured to justify inequality and oppression, and has no bearing on the everyday lived realities of those who are deemed incurable.

"Cure is inextricably linked to hope," scholar-activist Eli Clare brilliantly writes. It lies in the eradication and the violence that accompanies it, connects to both elimination of self and erasure of identity. It makes possible all that was impossible and unfathomable, delivers on demands but doesn't always distinguish between the real and the fake.[82] It only matters that it exists, this cure, and that it can deliver. If it fails to do so, then it's not the cure's fault. It's yours.

So much riding on a single word. Hope.

1

Improbable Miracles

Frank wanted me to tone up my body by tonic medicines and restore my hearing with almond oil, but, *prosit*, nothing came of the effort; my hearing grew worse and worse. . . . Then came a medical ass who advised me to take cold baths . . . prescribed pills for my stomach and a kind of tea for my ears. Since then I can say I am stronger and better; only my ears whistle and buzz continually, day and night. I can say I am living a wretched life.

Ludwig van Beethoven to Franz Gerhard Wegeler, 28 June 1801[1]

For Deafness, Singing in the Ears, &c.:
Take the juice of sow-thistle and heat it with a little oil of bitter almonds, in the shell of a pomegranate, and drop some of it into the ears. It is a good remedy for deafness, singing, and other disease of the head and ear.

Pierpont F. Bowker, 1851[2]

In 1854, a pamphlet circulated around Charleston, South Carolina. Anonymously written, the slim publication drew citizens' attention to the recent arrival of a Scottish physician whose advertisements for deaf and blind cures provided only slight clues to his mysterious identity.[3] Gossip mongers tittered over rumors of his medical prowess. Hushed whispers shared the tale of how not two years prior, First Lady Mrs. Abigail Fillmore had recommended the physician's services to her friend, Sarah Helen DeKroyft. A few weeks after her husband fell out of a carriage to his death on their wedding day, DeKroyft had woken up to find herself suddenly blind, after her tears caused an infection in her eye. She remained so for seven years, giving up on any oculist to cure her—until she met Dr. Turnbull, whose uncanny abilities to restore sight to the blind and sound to the deaf restored her faith.[4]

DeKroyft's doctor was not the same individual who arrived in Charleston. That fact did little to quell chatter about the enigmatic Dr. Alexander Turnbull. He had arrived in New York on a steamer from England, citing a necessary change of climate to revitalize his health. After touring the states with his two daughters, he took up lodgings in a Charleston hotel. Shortly afterward, he sent out advertisements announcing his ability to cure all cases of deafness and blindness. He assured citizens that his treatments were "simple and painless," offering scores of transnational testimonials from satisfied patients.

No one paid attention at first. After the *Charleston Courier* published accounts of two of Turnbull's cases, his fame began to spread. He had cured a nine-year-old boy from Georgia who had been completely deaf since age three, and an eighteen-year-old with an alleged hereditary hearing and speech disorder. Prior to visiting the foreign doctor, the children's parents had exhausted all possible medical avenues at considerable expense. They had given up the possibility that the youth's hearing could be restored.

Was it a miracle? Did Dr. Turnbull manage, despite all odds, to cure the incurable with his secret process? Or was he simply "making all sorts of noise," as educator Joseph Henry Johnson believed?[5] Johnson, who would receive his medical degree from Jefferson Medical College and establish the Alabama Institute for the Deaf, Dumb, and Blind—the first such institution in the state—was suspicious of the physician's intentions. "Dr. Turnbull from *Scotland* who operated on the Polls' Ears he pretends to cure cases of congenital deafness," Johnson wrote to his wife Emma.[6] As Johnson already knew, despite the protestations of medical men, most cases of congenital deafness were in fact simply incurable. His own brother, William Seaborn, was the only deaf child in a family of ten children.[7] No medical treatment, no matter how plausible, could restore Seaborn's hearing.

The anonymous writer of the pamphlet agreed with Johnson and declared that no miracle was performed by Turnbull. Rumors of Turnbull's abilities had been grossly exaggerated, fueled primarily by the misleading advertisements he placed. Upon investigating, the writer discovered that Turnbull's countrymen held him in low esteem, especially once suspicion arose that he was somehow involved in the death of a patient following a dangerous surgical procedure. There were more shocking revelations. Turnbull not only attempted to cure deaf-dumbness by an "electro-stimulation experiment," but claimed he succeeded with several deaf-mutes in Scotland and at the Deaf and Dumb Institution of Edinburgh. So numerous were his "unprofessional and disreputable" acts, the writer considered it a wearisome persecution to repeat them in his pamphlet.

Turnbull's worst offense, the writer emphasized, was his unblushing effrontery to prey upon anxious parents. He promised rapid, uninterrupted progress in hearing and articulation. He explained that—upon payment, of course—several courses were required to regain healthful use of dormant faculties. Each course required additional payments. He gave parents hope that a perfect cure awaited on the horizon. And these parents, the writer stressed, remained encouraged even after they departed Turnbull's office in trembling anxiety, only to eventually "have the unwelcome truth forced upon them, that the objects of their solicitude were as hopelessly incurable as ever." As the Georgia boy's parents ruefully averred, their son remained deaf despite Turnbull's insistence that he could hear.

It is easy to be swept away by promises of miraculous cures for incurable diseases.

Until the advent of the germ theory in the nineteenth century, the concept of the four humors dominated medical discourse and theories of disease transmission, configuring illness as a matter of inherited susceptibility and individual temperament. The body was a system of equilibrium: health was the product of proper balance within one's system and that of the environment; illness, the destabilization of the system. External influences could corrupt the body by inhibiting the proper release of bodily fluids—blood, urine, phlegm, bile—which then had to be manually drawn out.

When there were no obvious environmental factors to explain the cause of a disease, alternative explanations were given. Diseases were the invasion of demons, curses cast by those with malevolent intentions, or the result of an unsteady climate corrupted by tiny particles in the air. The aim of the medical practitioner was to devise a treatment to extract these trapped impurities from the body and restore it to its natural harmonious state. They offered plenty of options, all based on the same thing: to treat a sickness, one must purge it out. Drain copious amounts of blood; take emetics to induce vomiting; use laxatives to promote diarrhea; prompt sweating; take a trip to the countryside for a change of air; pray to the saints or take a pilgrimage to a holy shrine; purchase a carefully prepared dose from an apothecary. For any sickness, there was a way to purge it out of the body. This made medicine a risky business, for oftentimes it was uncertain whether the risk of cleansing impurities from the body would outweigh the dangers of death.

For the most part, sick people relied on self-treatment. Whether it was using tried-and-true family recipes, accepting the advice of sage neighbors, or purchasing household health manuals, there were plenty of options to avoid invasive interventions—especially since a healer's fee was often beyond the reach of many. This still holds true nowadays. I always begin my history of medicine course by asking students what the *first* thing is that they do when they are sick. The answers are diverse, but usually the same ones appear every semester: sleep, drink orange juice, complain, take an Advil, or ask Mom. Hardly anyone answers "see a doctor." That's because we seek a healer only when we don't know how to help ourselves or when we don't understand the root cause of our illness. And when we become frustrated with the limitations of professional medicine, like so many sick people before us, we tend to seek out alternative options for bringing our bodies and our selves back to health.

The precepts of self-medication brought egalitarianism to the art of healing, providing an alternative to the doctor's "bleed, blister, puke, and purge" gambit, which too often made the body sicker.[8] In America, as in Western Europe, domestic medicine provided powerful and efficacious means for healing. Relying on the curative properties of botanics to create medicinal preparations, domestic medicine allowed people to blend folk traditions, superstition, and faith with other medical methods—including homeopathy, vegetarianism, and fasting—to adopt an "all of the above" approach to healing or as a last resort where medical practitioners had failed.[9] Largely in the province of women, these preparations were verbally passed down through generations or written in receipt books alongside cookery recipes, physicians' advice, and extracts from household health manuals. While the bulk of these domestic remedies were based on common sense and tried techniques, they mostly alleviated symptoms rather than cured the patient.

And what of deafness? Those who immigrated to America from Europe brought with them an analogous assortment of domestic and folk remedies. There were recipes making use of ants' eggs mixed with onion juice, a longstanding solution from Scotland and popular in the old west.[10] Another recipe called for onion wrapped in tobacco leaves and baked, the juice squeezed out and poured into the ear. Sow-thistle juice, oil of bitter almonds, houseleek, sassafras oil, black radish, black wool, animal oil, and insects were all suitable as ingredients. Treatments were also borrowed from Native Americans, building on a system of polypharmacy by making use of various substances to treat a disease.[11] From the Blackfoot tribe came the idea of using wintergreen or the juice from serviceberry on the eardrums, as did the Tarahumara's use of the pincushion cactus.[12] From the Cherokee tribe as well as from African-American slaves, the use of garlic, a staple for thousands of years for its antibacterial and diuretic qualities, was recommended as a treatment for deafness when placed into the ear.[13] And some recipes were published for widespread dissemination. Jane Francesca Agnes, wife of aurist William Wilde and mother of Oscar Wilde, for instance, recorded an ancient Irish cure for deafness: "nothing is esteemed better than constant anointing with the oils of eels, used perfectly fresh."[14]

As historian Seth LeJacq points out, medical practitioners also invoked domestic treatments to balance the contours between botanical medicine and invasive surgery, proposing new theories and treatments in response to "patients' fear of surgery and desires to avoid pain and pursue gentle remedies."[15] Physicians merged critical elements of herbalism, astrology, and ritualism; many of these cures borrowed heavily from the work of Roman surgeon Galen or from Pliny the Elder's *Historia Naturalis*. They also introduced new therapies like hydropathy, phototherapy, mesmerism, galvanism, and radiotherapy, desirable options for those with debilitating illnesses and chronic conditions. Surgical intervention for deafness, of course, remained an option, though the theoretical basis for procedures remained speculative or experimental at best until the nineteenth century. Yet even as top aural specialists in Britain, France, and Germany advanced their knowledge on the anatomy and physiology of the ear, many practitioners admitted they found it difficult to diagnose a cause for deafness in order to prescribe an appropriate treatment course; in most instances, treatments not only were ineffective, but actually caused further damage to the ear.

Nevertheless, aural specialists generally agreed that the causes of deafness could be classified into two categories: "obstructed hearing" or "sudden hearing loss," the result of mundane conditions such as earwax buildup, damage to the eardrum, or disease; and "exhausted hearing" or "nervous deafness," a more permanent condition resulting from damage to the auditory nerve. Patient history and proper examination of symptoms were necessary for forming a correct diagnosis; acquired deafness, or hearing loss occurring later in life, was more common than hereditary deafness. Unfortunately, too often patients found themselves lumped into a third category: "total deafness," an incurable state with no chance of hearing restoration, even though specialists and quacks alike both claimed they could devise a cure.

As long as people worried about their hearing—and their health in general—there were sellers hawking miraculous remedies. The Jacksonian call for self-determination was particularly prevalent on the frontier and rural areas, where geographical isolation encouraged Americans to rely on instruction books and commercial

FIGURE 1.1. "Breathing down the fingers into the ear for deafness." From D. Younger, *The Magnetic and Botanic Family Physician, and Domestic Practice of Natural Medicine* (London: E. W. Allen, 1887). Wellcome Collection. Creative Commons Attribution (CC by 4.0).

medicine. "Oral and written traditions shaded into folk beliefs and superstitions," historian John C. Burnham tells us. "Everyone, from the most learned to the most superstitious, had 'recipes' for treating various ailments. Often, too, people improvised ways to prevent and cure disease."[16] The healing ministry of Protestantism, the suggestive guises of hypnotism, and the faith healers who openly disparaged doctors: they also provided alternative avenues, rooted in an old phenomenon, the use of laying hands on the sick, the calling on the divine to arrive from the heavens and penetrate the diseased body. Whether they are curious wonders or mere charlatans, their sects flourished and will continue to flourish to service those who are willing to line up, assert their faith, and pay.

Red Hot Chili Peppers

The chili pepper is a staple in many Punjabi dishes, used to accent or heighten flavors of *sabjis* (curry dishes). The kitchen tables of my childhood always contained centerpieces of jars with pickled chilis, mangoes, and onions. They were part of the dining experience, the contents serving to cleanse palates between various *sabjis* my parents had prepared. The *mirchi*, as the chili pepper was referred to in our household, was also used as punishment: when we children spoke out of turn or misbehaved appallingly, we were ordered to bite the *mirchi*. Even the mere threat of being forced to take a bite was enough for us to snap back to our docile selves. My association with the *mirchi* also has another powerful memory: the monthly or so ritual of my mom cleansing our home of the "evil eye."

Punjabi tradition identifies the *mirchi* as a powerful tool to ward off malicious thoughts and vile intentions from strangers. The envious glance from insecure humans and supernatural beings was perceived as damaging enough to leave behind a malevolent residue. Children, and especially infants, were believed to be the most vulnerable to the evil eye. Even the sudden occurrence of a disease indicated the presence of an evil eye, which could be corrected by protective and healing measures, often one that created an equally repugnant smoke to counteract the malevolent residue.[17] Like the one trailing from burning chilis. My mom would grasp several burning chilis, circle it around our heads several times to pick up the "sickness," then throw them into the stove. The crackling sound and irritating odor is evidence of the extracted sickness, the evil effect, being destroyed, completing the healing process.

Over the years, the ignited *mirchi* made many rotations around my head and through rooms in our home. Sometimes it possessed additional powers to ward off malicious intentions from strangers that were aimed at making my hearing progressively worse. The stinging, painful aggravation from the *mirchi* that seared my skin and watered my eyes was mere nuisance compared to the lifetime of protection and strength that it provided.

At the public library one day when I was ten or eleven, I came across a large book on Greek mythology. Reveling in the stories of the Olympian gods and goddesses, I learned how each of them ruled their own separate domains—for instance, Poseidon over the waters and Hades over the underworld—to invoke their powers towards humanity and the natural world. Ares, the violent, untamed, and fierce God of War, son of Zeus and Hera, held domain over all warrior energy. According to ancient lore, Ares's energy (or more broadly, that of his Roman equivalent, Mars) was transmitted to certain red foods, like the chili pepper, which I learned is associated with Ares's ruling planet as well. The association provides the pepper with passionate and energetic power that becomes valuable for breaking hexes, especially when burnt, as its irritating and fiery scent can quickly envelop a room. No wonder medieval folklore refers to burning chilis being used as deterrents against vampires and werewolves, or that witches and doctors alike used it to excise demons as well as humors. The scent of a burning chili is nearly impossible to forget.

The chili pepper is one of many vegetables believed to possess powerful medicinal and magical properties to aid the body in healing. Deriving from folk traditions and merged with rituals and superstitions, domestic medicine was—and is—the cornerstone of everyday health, relying mainly on the cultivation of botanicals and animal products. While domestic healers tended to document their preparations in leather-bound books or family bibles and pass their knowledge through oral traditions, physicians often compiled elaborate catalogs on the medicinal benefits of various botanicals.[18] Inheriting the legacy of the Greek botanist Dioscorides, early modern physicians had over seven hundred medicinal plants at their disposal for managing illness. Later herbalists, particularly the English physician Nicholas Culpeper, refined the botanist's works, expanding knowledge of medicinal herbs and developing more powerful prescriptions. Culpeper's influential catalog of herbs, *The English Physician*, first published in 1652, outlined how the uses of plants were closely intertwined with astrological readings. Such catalogs — known as "herbals" — were heavily relied upon by local healers as well as physicians like Culpeper. They referred to the medicinal properties of plants in conjunction with astrological readings to extract the maximum benefit of the plant's curative properties.

These manuals provided instructions on how to use plant and animal products in medicinal preparations. Edward Topsell's *The History of Four-Footed Beasts and Serpents*, a three-volume compilation published in 1658, contains stories of how to use animals and ancient cures to treat a variety of ailments and disorders. Topsell, an English clergyman, gives advice for situations where "one be deaf or thick of hearing."[19] "Take the gall of an Ox and the urine of a Goat," one passage advises. Or the "Gal of Goose." The same remedy, if applied to the temples, can ease headache tension. If applied to a woman's breasts, it can prevent her milk from curdling.[20]

If you happen to have a lion nearby, or are tired of your hateful cat, their feline brains "mingled with some small quantity of oil of spike," distilled into the ears of a deaf person, will cure them. So too, will "fat of mad dog," or calves' marrow mixed with whey, rosecake, an egg, and a dash of cumin. Wolf blood mixed with oil could work wonders too, and if such recipes were too complicated or ingredients too rare, goose grease poured into the ears was a proven remedy.[21]

Household health manuals also borrowed from these herbals, offering literate domestic healers the ability to draw from ancient healers. Many recipes overlapped with cookery; after all, remedies and meals required the same ingredients. Physician Hugh Smythson's popular manual *The Compleat Family Physician* (1785) outlines recipes for different "types" of deafness that could be made in the home kitchen.[22] *The Druggist's Hand-Book of Practical Receipts* by Thomas Branston, first published in 1853, lists a simple remedy for deafness that could easily be mistaken for a culinary recipe: "Oil of almonds, 1lb., bruised garlic, 2 oz., alkanet 1/2 oz; infuse and strain. Apply with a little cotton to the ear."[23] In rural America, where competent physicians were rare, these books became indispensable sources of treatment for many families. The more popular or comprehensive ones were even imported from overseas by wealthier citizens. These were optimistic books: every ailment afflicting the body had a cure of some sort.

Household recipe books also represented a "first port of call" for treating common ailments in domestic settings, serving as a means for (quickly) regaining health and avoiding unpleasant therapies or high fees from the physician.[24] These books include clipped or handwritten recipes copied faithfully by domestic healers—mostly women, though collecting recipes was an activity that interested men as well—who typically had the leisure time to collect, record, and compile these books from a variety of sources.[25] Ingredients were simple and easy to obtain. Herbs and other basic materials could be purchased from a market, plucked from a forest, or grown in a garden.

The notebook compiled by a Miss Barton of Suffolk from 1758 to 1766 includes a recipe to cure deafness: "Take some green wormwood & rub it in yr hands till it is very moist then put it in ye hollow of yr ear & it will cause [it] to Discharge you must repeat it every day or two."[26] Another home remedy book dated to 1896 copies a treatment for earache extracted from the popular American magazine *Farmer's Friend*: "A remedy which never fails is a pinch of black pepper gathered up in a bit of cotton batting wet in sweet oil and inserted in the ear. It will give immediate relief."[27] Some women published their medicinal and cookery recipe books to be used as reference manuals for others. In her *Modern Domestic Cookery*, Elizabeth Hammond includes a cure for deafness: "Take an ounce of the oil of sweet almonds, and the same quantity of camphorated spirits of wine mix them well together, pour a few drops into the ear every night, and put in a bit of cotton." She advised that whatever the cause of deafness, the head should "always be kept warm"; a tablespoon of bay salt in half a pint of spring water, soaked for twenty-four hours and poured into the ear for up to nine successive nights, was also an effective remedy for deafness.[28] An 1867 cookbook by Annabella Powell Hill (neé Dawson) also includes a remedy for deafness, which required dividing an onion, filling its center with a "fresh quid of tobacco," binding the onion to roast it, then juicing it.[29] Three drops were to be inserted into the ear, and though it could be painful upon first application, it was a guaranteed cure. As one of the most influential Reconstruction-era cookbooks, Hill's recipes ended up in areas as far north as Maine. These recipe books mainly treated minor complaints that did not require a physician's care. They also did little to discern between temporary and permanent afflictions of deafness, though it's evident that an "earache" was differentiated from "deafness." Manuals with advice borrowed from medical textbooks, however, did list solutions for deafness according to a suspected cause. For partial deafness, one recommended prescription read as follows: "Take eel-fat and mix it with the blood of an eel and goat. Put the mixture in a hollowed-out onion and place on the fire until the onion begins to soften. Then take the onion and squeeze two drops of the liquid into the ear."[30] An 1840 treatise compiled from various American and English publications lists several remedies for deafness: For a buildup of wax, regular syringing and washing with warm soap and water was recommended. A tincture of myrrh and Egyptian honey was useful for deafness caused by ulcers. When caused by a "decay of the nerve," electric sparks, a blister behind the ear, or sneezing powder were the most powerful remedies.[31]

While these books were valuable sources of medicinal knowledge, they could be tricky sites of misinformation, outlining ineffective, if not downright dangerous, advice. The English aural surgeon William Wright, a popular and prolific writer, worried that ineffective home remedies would further damage deafened ears and thus discourage people from seeking out qualified advice when required. This could especially be the case with poor people, who "either resort to the use of some of the numerous infallible remedies with many cookery and 'receipt' books, or old women's brains . . . purchased the advertised nostrums of the day, or become the dupes of an ignorant empiric."[32] Wright argued that "fashionable" ancient remedies such as those cited in the sixteenth-century manual *Thousand Notable Things*—and reprinted in pamphlets and medical textbooks—were mostly worthless, especially those that included animal substances. Fat of eel, cast skin of serpent boiled in wine, ant's eggs, wood lice heated in rose oil, fat of cow's feet, fat of fox lungs, or goose or duck grease: these did nothing to alleviate deafness. Wright encouraged deaf persons to take careful stock of advertised remedies that appeared fanciful but were in fact dangerous and useless. These included corrosive sublimate dissolved in lime water, any form of mercurial drops, or galvanism treatments that were often presented as infallible cures by incompetent charlatans.

Before Alexander Turnbull arrived in Charleston, for instance, he was well known in Scotland and England for advocating the use of veratria, a poisonous alkaloid obtained from the hellebore root. When ground into an ointment and applied to the external ear, Turnbull claimed it could cure all cases of deafness. Alkaloids were commonly used in medical mixtures during the nineteenth century, and although some were proven useful in topical remedies, physicians were doubtful of their efficacy. Combined with other alkaloids from plants in the Ranunculaceæ family, veratria, Turnbull boasted, was effective for other conditions as well, including gout, dropsy, rheumatism, neuralgia, and heart conditions.[33] In advertisements, he referred to his veratria remedy as "electro-stimulation treatment," even though no electrical charge was ever emitted.

Dieting Deafness Away

In his humoral system of medicine, the Greek physician Hippocrates of Kos emphasized the importance of diet and exercise. Eating the right foods and moving the muscles were necessary for expelling toxins from the system to maintain humoral equilibrium. Certain animal products and plants—like chilis—could aid in that process, for they possessed powerful elements for boosting health or dispelling illness. As medicinal cookery provided solutions for ailments, diet was devised as a form of self-medicament, even advocated by physicians to help their patients regain their natural equilibrium.

William Banting is one individual who attempted to diet his deafness away. Banting was an upper middle-class funeral director whose family held the Royal Warrant for burials for five generations, until 1928. George III, George IV, Prince Albert, and Queen Victoria were all buried by the Banting family. At sixty-six years old, Banting was retired and a widower. He was also obese, 5′5″ and 202 pounds at his highest weight, and wore a truss to hold in place an umbilical rupture. His weight was so troublesome that he could not tie his own shoes and had to go down the stairs backwards to limit the pain of weight on his knees and ankles. He had tried several methods of dropping pounds, including "sea air and bathing in various localities, with much walking exercise; taken gallons of physic and liquor potasse, advisedly and abundantly; riding on horseback; the waters and climate of Leamington many times, as well as those of Cheltenham and Harrowgate frequently."[34] These attempts were in vain.

In 1862, after finding his eyesight failing and his hearing greatly impaired, Banting consulted an aural surgeon. He was disappointed, however, by the surgeon's approach, for he "made light of the case, looked into my ears, sponged them internally, and blistered the outside, without the slightest benefit, neither inquiring into any of my bodily ailments, which he probably thought unnecessary, nor affording me time to name them."[35] After the surgeon left on holiday, Banting decided to visit the Royal Dispensary for Diseases of the Ear (est.

1816), which was managed by William Harvey. A respected aural surgeon, Harvey was educated at London's Guy Hospital and practiced general surgery for several years before deciding to specialize in ear diseases. In addition to his post at the Royal Dispensary, he was aural surgeon at the Freemason's Asylum for Female Children and at the Great Northern Hospital. As he repeatedly stressed, he bore no relation to the famous physician William Harvey, who experimented with blood circulation.

Upon examining Banting's ears and assessing his obese stature, Harvey concluded that the undertaker's deafness was a symptom of fat deposits lodged in the Eustachian tubes. Recalling how starch could negatively affect the body and its role in the production of insulin, Harvey advised Banting to completely overhaul his diet to lose weight, explaining that as the pounds dropped, his hearing would be improved. Banting was advised to abstain from bread, butter, sugar, beer, and potatoes, which had "been the main . . . elements of [his] existence." Instead, he was instructed to eat four meals a day composed of meat, greens, fruit, and dry wine. The diet worked: Banting lost forty-six pounds in a few months. He was so astonished by the seemingly miraculous results that he paid Harvey an extra fifty pounds on top of his usual fees, to be distributed among his favorite hospitals.

Unfortunately for Harvey, false rumors spread in London that the diet had ruined Banting's health and that the surgeon was to blame. Furthermore, Harvey's practice suffered as he was constantly ridiculed for being unable to explain the connection between the diet and Banting's improved hearing. Since Harvey's diet management could not be replicated under similar variables, medical practitioners criticized Harvey's claims that the diet was "scientific." It was preposterous to even suggest that deafness could be dieted away!

Meanwhile, Banting found himself attaining celebrity-like attention after he self-published a pamphlet in the form of an open testimonial in 1863, *Letter on Corpulence, Addressed to the Public.* The pamphlet sold out its initial printing; later editions were published by Harrison of London and copies were eventually picked up by American publishers, disseminating Banting's celebrity to new lands. Banting's dietary success—in losing weight, more than regaining his hearing—became so well-known that asking someone "do you bant?" was a fashionable way of asking whether people were dieting.

Banting's diet remains ingrained in our culture. We just call it the low-carb diet.

Other dietary systems attempting to alleviate deafness had their moments. The editor of *The Rural New Yorker* undertook a diet of pork based on his landlord's theory that the ears of the hog are exceedingly acute and could be passed on to the meat. The diet did not cure the editor's deafness, but he surely enjoyed four months of fresh and salt pork.

William Oakley Coffee of Des Moines, Iowa, appeared on the advertising circuit during the early nineteenth century for his "Elimination-Absorption Treatment" for deafness and ear diseases. Claiming a diploma from Missouri Medical College in 1881, Coffee initially practiced as a general physician before beginning a mail-order "eye-cure" and "spectacle-fitting" business that was eventually exposed by writer Samuel Hopkins Adams in *Collier's* "Great American Fraud" series that ran from 1905 to 1907.[36] Filing for voluntary bankruptcy shortly after the exposé, Coffee carried on his itinerant practice before settling in Davenport and establishing a new mail-order business of "deafness cures." Taking out full-page advertisements costing as much as $190,000 a year, Coffee's success grew exponentially; he even purchased valuable "sucker lists" (names of supposed sufferers from deafness and catarrh) from letter brokers for more targeted advertising. And within the Midwest, his advertisements "hit upon a very lucrative field—among the Ku Kluxers."[37]

Sixty thousand people used his treatment, Coffee boasted. It began with a complete reworking of diet to cleanse the stomach, liver, bowels, kidneys, and blood, for "you cannot build up a good, strong, healthy tissue over diseased, dead waste matter."[38] Dieting, Coffee explained, was particularly important in the treatment of catarrhal deafness (when mucus blocks the ear and throat cavities) and other forms of partial deafness, for deposits from the nose and throat are oftentimes swallowed and retained in the stomach and bowels, causing discomfort and an unhealthy constitution. This was the "secret" of why Coffee's treatment worked where others had failed, a claim affirmed by the "testimonials" in his advertisements. Following diet, the "Full Treatment" is then offered: a massage of the nose and Eustachian tubes, following by a cleansing of the aural system using "healing vapors" injected into the nose with a syringe. Then, an antiseptic absorbing balm is applied to the membranes of the nose and left to dry. Instant success.

"Now I Hear Everything! Say Many Fortunate People."[39] The United States Post Office disagreed; in June 1935, it issued an order against the Coffee Company (then managed by William's son Percy after the elder Coffee's death in 1927 from a heart attack) for their mail-order scheme conducted through false and fraudulent pretenses.

Another dietary miracle for deafness appeared around the same period. In 1934, Dr. Grant Selfridge, a graduate of the Hahnemann Medical College in San Francisco and member of the California State Homeopathic Medical Society, examined the medical histories and nutrition diaries of his deaf patients.[40] He observed that most of his patients were deficient in vitamin B, which was essential for maintaining healthy nerves. Selfridge became convinced that vitamin B was a contributing factor in hearing loss, easily remediable through diet, especially for patients with otosclerosis, a hereditary condition that causes progressive deafness due to bone overgrowth in the inner ear. In over a hundred patients, Selfridge prescribed vitamin B tablets, rice brans, or injections. After as few as six injections, he observed evidence of improvement, especially in younger patients; older patients required longer courses of treatment.

Five years later, Selfridge achieved national fame when *Time* magazine published a short article on his use of vitamin B to treat deafness caused by deterioration of the eighth cranial nerve (the

auditory vestibular nerve, which transmits sound and balance information from the inner ear to the brain). After the Associated Press
picked up the story, a flood of inquiries and patients from all parts of
the country arrived in Selfridge's office. In a later publication, however, Selfridge clarified that further research was required to study
the entire vitamin B complex and its relation to healthy nerves. By
1941, he was experimenting on rats, guinea pigs, and chicks that had
been fed diets deficient in one of the components of the vitamin B
complex.[41] His research led him to advocate that pregnant mothers
should add more vitamins to their diets as a preventive measure
against deafness and other disorders: "The mother should have an
optimum diet during her entire pregnancy, and the child when born
should be treated the same way. The dietary correction should be
carried through childhood and adolescence. This is the only way the
deafness can be obviated.... The important factors in diet should be:
One egg daily; at least one pint of milk daily; one leafy salad daily;
orange, grapefruit, or lemon daily."[42]

Does an increase in vitamin B cure deafness? The 1930s and 1940s
featured numerous studies on vitamin and nutrition therapy for
deafness alongside Selfridge's, but none conclusively proved that
deficiency of any vitamin can cause deafness. Recent anecdotal research suggests that taking vitamin B may improve tinnitus ("ringing in the ear" sensation). Some have examined the effects of vitamin B-12 on hearing loss, arguing that deficiency in the vitamin can
lead to, or worsen, cases of progressive hearing loss; however, similar studies have argued the contrary.

A quick internet search reveals a host of advice for "natural healing" to prevent deafness. Lists with titles such as "eat better, prevent
hearing loss" include food containing essential vitamins and minerals supposedly beneficial for the auditory system: zinc, omega-3
fatty acids, vitamin A, vitamin B, calcium, magnesium, and vitamin
K-2 are all on the list. Some websites additionally provide dietary
guidelines to prevent hearing deterioration, though there is insufficient evidence to support any of these claims.

Inside the Labyrinth

"Diseases of the ear have been for a long time the opprobrium of surgery," wrote G. Grant of Newark in 1857. "The hesitancy of the general practitioner," he continued, "to interfere in cases of simple deafness, to investigate his patient as to whether his case is one of structural or functional change, too often causes the patient to despair all together [*sic*] or throw himself onto the hands of the advertising quack."[43] Grant's words reflect the consensus among physicians of the period who were unfamiliar or untrained in the intricate aspects of the ear: attempting a cure for deafness could be a pointless endeavor.

American physicians may have played a crucial role in winning their nation's independence, but they lagged far behind their European counterparts when it came to innovative and inventive medical advancement. A handful of colleges were built across eastern states, but American medical infrastructure was comparatively underdeveloped. A physician's career depended on house calls, but home remedies and self-treatment remained the cornerstone of health care, particularly in rural areas where access to a physician was difficult. Eager to avoid implications of cultural and intellectual inferiority, and anxious to obtain the clinical opportunities of Paris and laboratories of German universities, (wealthy) medical students flocked to the Continent to learn the tools of their trade.[44] They also sought to develop a clinical specialty that would give them an advantage within the American health market. Cardiovascular diseases and obstetrics were popular choices. The ear? Hardly a lucrative option to build a career on.

The ear is a complex organ. It is divided into three parts: the external, the middle, and the inner. What we refer to as "the ear" is really the external auricle, furrows and depressions of flaps of cartilage and skin serving to gather and funnel sound inside the ear canal. The ear canal itself is about an inch and a quarter in length, surrounded by hair, which prevents foreign bodies from entering; wax, secreted by skin glands, also traps extraneous substances. The middle ear begins where the ear canal ends at the tough, angular tympanic membrane, also known as the eardrum. Inside the middle ear, filled with air and enclosed in the bony structure of the skull, reside the tiny ossicle bones (the incus, malleus, and stapes), which transmit sounds to the inner ear. The middle ear also connects to the throat through the Eustachian tube, a passage that opens when you yawn or swallow, its function being to provide ventilation and drainage for the middle ear and to equalize pressure on the two sides of the eardrum. Finally, the inner ear, which is composed of the cochlea and system of semicircular canals, is the seat of your sense of balance. The cochlea, a snail-like coil of tissue and bone, contains the microscopic hair-endings of the hearing nerve; approximately 12,000 of them lie in the tissue in three rows; they amplify sounds and sharpen tuning. Another three or four thousand inner hair cells are tasked with sending signals to the auditory nerve fibers towards the brain.

I was eight years old when I first learned about the anatomy of my ear. My school had two classes for hard of hearing students, an elementary junior class, and a senior class for those in grades six to eight. I skipped grade five and was placed in Mrs. Ower's senior class, and because I was the youngest, I automatically assumed the position of my classmates' "kid sister." Until her death from complications caused by Parkinson's disease, Mrs. Ower was the guiding hand in my life well into my college years. It was she who first taught me the anatomy of the ear, teaching me how physiological and structural damage led to hearing loss. She explained that my ears didn't work because the "tiny hairs" deep inside my ears were damaged or lost. I listened, looked at the anatomy drawings she showed me, tracing the spirals of the cochlea with my finger, but I failed to understand her explanation. If hair grew everywhere on my body, then why didn't it continue to grow inside my ears? When I hit puberty and hair grew on parts of my body where there were none before, I imagined the possibility of new hair cells growing inside my ears. If they grew, could my hearing return?

We know now that there are three general types of hearing loss: conductive, sensorineural, or mixed. Conductive hearing loss occurs when there is a mechanical problem with the external or middle ear, usually the consequence of trauma: a serious injury or blow to the skull, excessive earwax buildup, ear infections, punctured eardrum, fluid buildup, or abnormal bone growth can all interfere with sound transmission. Environmental factors can also result in hearing loss, such as changes in air pressure or exposure to loud sounds, especially in enclosed spaces for prolonged periods of time, as in the case of boiler makers' deafness. Subacute infections — colds, catarrh, and other sinus troubles — can cause temporary deafness (or in rare cases, permanent deafness). Sensorineural hearing loss is permanent in nearly all cases. It has either a hereditary or a pregnancy-related cause (e.g., rubella or syphilis), or from childhood illnesses such as meningitis, scarlet fever, mumps, measles, diseases of the spinal cord, typhoid fever, herpes, tuberculosis, or Ménière's disease.[45] Trauma or abrupt changes in ear pressure can also damage the inner ear. Mixed hearing loss commonly occurs after extreme trauma or gradually over time.

In the nineteenth century, however, there was no consensus among medical practitioners about how to classify or treat hearing loss. Indeed, the aurist William Wilde described his medical colleagues as being "generally vague" on the subject of aural disease. "Until the greater questions in the physiology and anatomy of the labyrinth [of the ear] are positively settled," agreed American otologist D. B. St. John Roosa in 1885, "we cannot be sure of our classification of disease." With increased clinical and pathological study, he continued, it would eventually be possible for otologists to identify aural diseases with "as much accuracy as diseases of the heart, lungs or kidneys."[46] In the meantime, diagnosis relied on close inspection of the ear to determine structural abnormalities, and on a careful assessment of the patient's subjective symptoms to determine the degree of hearing acuity. The audiometer, an instrument emitting pure tones for identifying hearing threshold levels, was available as early as 1878, but most aural practitioners did not consider it a viable diagnostic tool, preferring instead to rely on conversational voice, speculums, ticking watches, or tuning forks to test for hearing loss.[47] These instruments weren't always reliable, which meant it was easy to misdiagnose a patient as permanently deaf—or "incurable," in nineteenth-century medical parlance.

Classification of aural diseases required a careful study of the symptoms to adequately prescribe treatment. Wilde, for instance, described inflammation of the external ear (acute erysipelas of the auricle) as expressing the following symptoms: redness, swelling, stinging, burning, headache, nausea, and fever. The recommended treatment relied on domestic healing and included a mixture of tonic purgatives of aloes, quinine, a nutritious diet with extra rhubarb, gentle exercise, and warm baths. Aggressive cases could benefit from a mercury concoction, though not all agreed on its benefits. William Wright, for instance, argued mercury was useless, if not outright dangerous, for such minor inflammations. Most cases encountered by aurists were of the noncongenital type and readily curable, mainly by syringing to cleanse or remove wax and/or foreign bodies from the ear, nose, and throat. Poultices were helpful in cases of extreme pain, especially with the addition of opium or morphine inserted into the ear canal. In cases where the eardrum appeared damaged, bloodletting or leeches could help remove pressure from the arteries. Surgical removal of tonsils or polyps could improve cases of hearing loss due to blockages.

Congenital cases were tricky. Although some aural practitioners knew that their deaf patients were unaware of their hearing loss (having never heard sounds before), they struggled to explain the etiology of congenital sensorineural deafness.[48] Some practitioners offered creative explanations blaming modern life: alcoholism in the parent(s), quinine poisoning, falls into the water, lightning strikes, tobacco use, and sunstroke.[49] They also offered superstitious explanations, as in the case of a ten-year-old girl who was frightened by her classmates to the point her head ached; the next day she became deaf, and even three months later, after being treated by the eminent surgeon Sir Astley Cooper, she continued "in the deplorable state of deafness."[50] Poor hygiene and living choices, including improper meals, lack of exercise, or improper rest, could also result in deafness.[51] Many of these cases, however, remained incurable, but they were also rare cases. The Richmond Eye, Ear, and Throat Infirmary's 1888 report, for instance, outlined 939 cases of ear diseases. Of these, 606 were conditions relating to the middle ear and treatable, including cases of aural catarrh, otitis, and perforated ear drums. Only 7 cases were of deaf-mutes; another 7 were permanent cases of deafness complicated by misuse of quinine, and yet another 7 were deafness caused by mumps.

For educators of the deaf, however, the incurability of (congenital) deafness was proof that deafness was a communication defect and that deaf people could "improve" through proper education based upon "moral management." For if deafness was incurable, then what grounds were there for medical practitioners to subject deaf children to painful treatments with no return? No treatment, whether syringing, herbal concoctions, blistering, or douching worked for these cases. Sign language and lipreading training, however, would at least give these children the skills to navigate through the hearing world.

Aural practitioners, of course, did not agree with the educators. While educators contested medical jurisdictions, the aural surgeon James Keene worried that this "very prevalent belief in the incurable nature of deafness" would become the principal cause for the medical profession's general neglect of the field.[52] And if there was neglect, there would be no advancement in the anatomy or physiology of ear diseases, which meant no new treatment to address incurable cases. Other practitioners, however, argued that educators and aurists needed to work together to improve the lot of deaf people. As early as 1810, William Wright and John Harrison Curtis appealed to London's educational institutions to allow them access to deaf pupils, insisting that a large percentage of the pupils had treatable cases of deafness. Without proper care, these children were at risk of permanent deafness—which, Curtis explained, could derail the efforts of educators to integrate deaf children into a hearing society.[53]

Certain childhood diseases, for instance, if left unchecked or misdiagnosed, placed a child at greater risk of permanent deafness, especially if the illness struck before the acquisition of speech. Diseases like mumps, measles, scarlet fever, herpes, and smallpox all threatened a child's hearing and were likely to proliferate in crowded quarters. Other diseases could be managed if caught at an early stage; syphilis, for instance, caused hearing loss and could be transmitted to the fetus before or during birth from an infected mother.[54]

Then there's the disease that violently tried to take my own life but settled for my hearing instead.

Once I was old enough to understand, my mother confided the reason I lost my hearing. I knew that a high fever consumed me and that I was sick for a long time, but I never knew the reason why, nor did I ever inquire. Perhaps I was afraid, or perhaps I didn't think this was something worth knowing. At my hard of hearing class at age ten or eleven, we were tasked with telling our life story. One question required us to share how we lost our hearing and the impact it had on our lives. It was probably the first time I asked my mother what happened to me. Her answer pinpointed a strange word, one that I could barely pronounce: "meningitis." "It's a disease that affected your brain," she elaborated. "Made you really sick with a high fever. You were so feverish that you kept vomiting and we were worried about you." In my memories, the moment my life was overtaken by fever began in a little room, or perhaps just a bed. Voices of my mother, father, and nanny came and went. The smell of vomit remained pungent. In my battle with the fever, nothing remained constant but the light that emitted from a green bulb that my father thought would be therapeutic for me. My memory stubbornly clings to the green.

According to a 2006 study by the Gallaudet Research Institute, bacterial meningitis is the second most common cause of hearing loss onset after birth, affecting approximately 3.2 percent of American youth. Meningitis is an inflammation of brain membranes caused by microorganisms penetrating the body through close personal contact (coughing, sneezing), a nasal infection, or even a skull injury. Viruses are the most common cause of the disease, but bacterial meningitis is deadly, especially if the bacterial strain is difficult to identify.[55] Early symptoms are usually mistaken for a cold or flu—as they were in my case—and the disease can quickly become life threatening. For those who survive the disease, complications can be severe: hearing loss, brain damage, kidney failure, and seizures. Today, vaccines are available for most strains of bacterial meningitis, but in certain countries the disease is still highly virulent and dangerous.

When seventeenth- and eighteenth-century physicians observed patients exhibiting symptoms of headache, high fever, and delirium, they formed a diagnosis of brain fever. The diagnosis went by other names too: "phrenitis," "cephalitis," "acute hydrocephalus," or simply "dropsy in the brain."[56] English physician Thomas Willis described the condition of "phrensy," noting that symptoms included "a deprivation of the chief faculties of the brain, arising from an inflammation of the meninges with a continual fever."[57] Epidemic outbreaks of disease enabled physicians to track the disease progression and chart symptoms, as Swiss physician Gaspard Vieusseux did in January 1805, when an outbreak hit Geneva and took thirty-three lives over a three-month period. Though Vieusseux, like other physicians of his day, could do little to halt the spread of the disease, the standard treatment was to cure observable symptoms, or at least prevent further progression of them. Phlebotomy, or bloodletting, was still a remedy of first recourse in Vieusseux's time. Other traditional remedies were also imposed, including mercury to clean out the stomach to remove excessive vomit, or sweating to perspire the disease out of the body. Bleeding, diuretics, and purgatives: the holy trinity of the medical armamentarium.

By the end of the nineteenth century, ear specialists were aware that meningitis was closely associated with deafness, as were school administrators who kept medical records of their pupils. The Illinois School for the Deaf, for instance, reported meningitis as the cause in 333 students enrolled from 1846–1896.[58] By 1887, Anton Weichselbaum of Vienna had isolated the causative bacteria of meningitis. Examining the spinal fluid of people sick during an epidemic, he identified the bacteria as meningococcus, which is one of the commonest strains of the disease. A new trend then emerged to treat meningitis: remove cerebrospinal fluid to drain the infection from the body. It would take another thirty years before the first effective medicine was used to kill the bacteria (sulfonamides), and another fifteen years after that for penicillin to prove effective before the introduction of vaccines.

My nine-year-old self was convinced that meningitis was an obscure parasite for me to vanquish, an invisible enemy that I needed to meet on the battlefield. Defeat the enemy, return home to glory. When my class shared the results of our assignment, I was excited over finding comrades to join the battle. Disappointment sunk in upon realizing I was the only one with this disease—the others had either been born hearing impaired or became so shortly after birth; a few had no idea what happened. Suddenly, I was alone on the field, with no one to relate to the excruciating realization that I must have done something wrong to get myself sick enough to be different from the rest.

Drumming the Ear

Seeking to transform the social and medical prejudice against the curability of deafness (and to make a name for themselves), some practitioners oscillated between offering surgical intervention and letting a case resolve itself. The physiological complexity of the ear, in conjunction with a general ignorance of the physics of sound transmission, meant that physicians too often neglected to properly examine a patient's ear and provide a competent diagnosis—either because they lacked the instruments or skills to do so, or because of indifference, knowing that there was little chance of improvement. Sir Astley Cooper, surgeon at Guy's Hospital in London, for instance, asserted that the ear is "too delicate an organ to be operated on."[59] If there was anyone who understood the difficulty of surgically treating deafness, it was Cooper. Of course, this did not stop him from devising a treatment.

In 1801 Cooper presented a paper to the Royal Society of London, a learned society whose members gather to discuss new scientific advancements. Discussing cases of hearing loss where he observed excessive wax or suppuration from infections accumulated in the internal ear, Cooper hypothesized that the buildup was blocking free passage of air through the eardrum, thus limiting hearing. Since the eardrum required air on both sides of the middle and internal ear to function properly, Cooper argued air was necessary for vibration against the eardrum for the transmission of sounds. A blockage meant no sound was being transmitted to resonate against the drum. Hence, no hearing. Cooper then devised a solution to this condition: puncture the eardrum. The collection of fluid in the middle ear would be drained and air would be released through the Eustachian tubes. If the puncture was properly done, damage would be minimal, and hearing fully restored.

This procedure, "tympanic membrane perforation" (now called "myringotomy") became one of the first to surgically restore hearing. Physicians marveled at the ingenuity of perforation and its advancement over traditional pharmacopeia, increasing optimism in the profession and in patients that deafness was surgically curable It also solidified Cooper's professional reputation. A year after he introduced his method, he received the Copley Medal, the Royal Society's oldest and most prestigious award, and became a fellow of the society in 1805, a remarkable accomplishment for any medical practitioner.

Cooper's insistence that perforation was beneficial only for a handful of patients—i.e., those whose deafness was the result of fluid buildup—was mostly ignored. It was more exciting to promote the treatment as a surgical remedy for *all* deaf persons than it was to lay out guidelines that excluded most patients. The frustration of treating obscure cases of hearing loss, including "nervous deafness," a condition with varying causes and symptoms, led even physicians

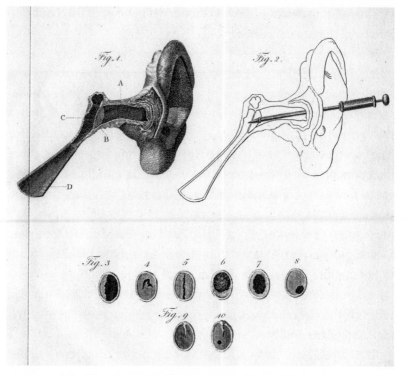

FIGURE 1.2. Astley Cooper's procedure for tympanic membrane perforation, 1801. From "Farther Observations on the Effects Which Take Place from the Destruction of the Membrana Tympani of the Ear: An Account of an Operation for the Removal of a Particular Species of Deafness," *Philosophical Transactions* 91: 435–450. Courtesy of Robert Ruben.

to unnecessarily apply the procedure across a broad spectrum of patients. Christian Michaelis, a professor of anatomy and surgery in Marburg, for instance, performed the procedure on sixty-three patients in 1804 alone, a tremendous number given that Cooper himself performed the procedure on a handful of patients during his entire career.[60]

The overreliance on perforation ended up increasing skepticism among the profession as to whether Cooper's method was nothing more than a lucky maneuver. As one writer commented in 1823, "London aurists I understand are, or pretend to be, of opinion that the drum of the ear requires a degree of opacity which is necessary to remove: this has been drummed into the ears of patients too much for me to hope to eradicate all at once."[61] Ridiculed for its proclamations as a permanent cure and criticized for its dangerous misapplication, perforation was perceived as a desperate attempt on the part of aurists to prove the worthiness of their specialty and to counteract cultural prejudice against the curability of deafness. Critics even argued it was imprudent for aurists to consider that such an invasive procedure could replace the standard medical armamentarium or centuries of botanical healing. Nearly twenty years after its introduction, perforation fell out of favor.

Nevertheless, perforation marked the beginning of a pattern in surgical cures for deafness: an innovative procedure is introduced and heralded as a miraculous breakthrough. Then, despite caution by the innovator that it should be applied only with a limited range of patients who met specific diagnostic criteria, the procedure is grossly misapplied to the point that it is rebranded as ineffective, falls out of favor, and is banished to the realm of quack medicine.

"If the cure of deafness," explained James Keene, "is not as frequent a result of treatment as we can desire, much of the blame is due to the patients themselves."[62] Too often, he continued, patients relied on home remedies, hoping to alleviate their deafness, and delayed seeing a specialist; then, anxious and desperate for an immediate cure, they demand a solution, jumping from practitioner to practitioner, from treatment to treatment, until they find something that could help. In some instances, their endeavors led to disastrous consequences.

Sometime during the 1830s, Alexander Turnbull ceased to promote veratria as the ultimate cure-all for deafness. He had left Hull to establish a practice near the exquisite gardens of Russell Square in London, and already he was facing the wrath of public and professional criticism over his medical qualifications. A diploma from Edinburgh's famed medical school, which he received in 1821, did little to quell popular opinion. In Hull, he came under fire after he was named the beneficiary of a large fortune left in the will of a patient, Mr. William Stephenson of Beverly. Hushed whispers suggesting Turnbull manipulated Stephenson to rewrite his will eventually turned into a full-blown scandal that reached the King's Bench in 1831. The foreman's reading of the verdict could not improve Turn-bull's faltering reputation: "We beg to be permitted to express our opinion, that Dr. Turnbull's conduct has been highly unprofessional and greatly discreditable to himself."[63]

Eight years later, Turnbull found himself in the center of another scandal. Two of his patients were dead.

Aurists were hardly surprised about the mess Turnbull found him-self in. For years, they had cautioned deaf persons not to be swayed by the boastful pretensions of advertised cures. British anatomist Joseph Toynbee, whose renowned dissections occupy a permanent place in the Hunterian Museum, castigated Turnbull's surgical skills. "Almost every medical man," he explained, "must have heard of the horrible effects sometimes produced by the application Dr. Turnbull uses. . . . It must be apparent that Dr. Turnbull has no greater knowl-edge upon the disease of the ear, than the ignorant." William Wright also warned the public not to be swayed by Turnbull's grand claims or his "extraordinary exhibition" of patients who were "successfully" cured of their deafness through a hundred-year-old surgical proce-dure: Eustachian tube catheterization.

Named after Bartholomeus Eustachius, who published the first detailed anatomical description in 1562, the Eustachian tube has long been known to play an important part in hearing.[64] Comparing the tube's structure to a quill pen, Eustachius believed its function was to provide another avenue for breathing, rather than assisting in the mechanism of sound transmission. Nearly 120 years later the French anatomist Guichard Joseph DuVerney corrected this mis-conception, arguing that the Eustachian tube was a means for re-newing air within the tympanum and for equalizing air pressure. He further declared that the tube is always open and that the eardrum maintained the flow of air, a notion Toynbee would later disprove.

Anatomical studies of the Eustachian tube revealed that the tube could be manipulated to diagnose and treat ear disorders. One of the most famous methods for restoring hearing through manipulation of the Eustachian tube was described by the surgeon Antonio Maria Valsalva in 1704. Known as the Valsalva maneuver, this manipulation requires a person to forcibly expire air through the Eustachian tube while holding their mouth and nostrils closed. This results in increased pressure to the tympanic membrane, and thus forces out blockages from the ear. Valsalva's maneuver dates to the Arabic physicians of the eleventh century who prescribed forced expiration of air as a remedy for cleaning out middle ear pus.[65] Nowadays, the maneuver is helpful for airplane ear, the moderate discomfort and pressure felt in the ears when a plane changes altitude during its descent. It also remains a staple in modern medicine, having been adapted as a diagnostic procedure for assessing hearing murmurs and heart failure.

Several modifications were made to Valsalva's maneuver during the eighteenth century to expand its range of application, since the method contained little therapeutic benefit for hearing loss not caused by blockages in the tube or tympanum. In 1724, Edmé-Gilles Guyot, a postmaster at Versailles, constructed an angular tube of pewter and put it through his mouth into the opening of the Eustachian tube. The instrument was then attached to a leather tube affixed to two small pumps that forced fluid into the pharynx and essentially "washed out" Guyot's deafness. Excited with his discovery of a newfound cure, Guyot presented his apparatus to the Royal Academy of Science in Paris, but it was not well-received. Anatomists did not think that the entire Eustachian tube could be reached through insertion.

Furthermore, physicians considered the treatment to be too cumbersome for regular practice. In 1741, Scottish physician Archibald Cleland, unaware of Guyot's apparatus, created a similar device for catheterization, but used a flexible silver tube, claiming that the material was suitable for handling the force of fluid through the pump. Like Guyot's invention, Cleland's met resistance from the medical community and was not an immediate success.

Despite the catheter's potential for revolutionizing therapeutics, physicians were reluctant to attempt on patients what they perceived to be a frivolous and dangerous procedure. Some simply declared that they lacked the dexterity to perform catheterization. It would not be until the 1830s, just as Turnbull was arriving on the scene, that catheterization of the Eustachian tube began gaining traction as a therapeutic application. Its acceptance was owed to French surgeons Jean-Marc Gaspard Itard and Nicolas Deleau, whose work at the National Institution for Deaf-Mutes in Paris encouraged more experimental surgical methods for treating deafness.

Experimenting on pupils at the institution, Itard was constantly in search of new methods and medicines, as if he were desperate to solve the puzzle of deafness.[66] He devised a wine-based therapy that included rose leaves, horseradish, parsley, sea salt, and pulverized asarabacca applied into the ear canal; the therapy was a failure. Eventually, Itard contended catheterization was the most beneficial method of treating deafness, modifying the catherization process by using solutions combined with salt or gaseous vaporizations such as tobacco smoke, coffee fumes, or ether.[67] Since the Eustachian tube connected the ear with the mouth—and thus, hearing with speech—Itard explained any blockages in the tubes severely interrupted the communication process. Forcing out blockages through catheterization would restore the natural link between the ear and mouth, and, with proper rehabilitation, a deaf child would thus be able to hear and speak.

In effect, with Itard's method of catheterization, deafness could be cured, as it was for twelve-year-old Christian Dietz, a deaf-mute from birth and one of Itard's most famous cases, secondary only to Victor, the "Wild Boy of Aveyron."[68] According to reports, Dietz's hearing was fully restored after Itard's catheterization and he even recovered his speech. However, the same procedure failed to remedy hearing loss in twelve other pupils at the institution, leading Deleau to introduce his own modification to the senior surgeon's procedure. Catheterization with an injection of fluid, as was Itard's preferred method, Deleau argued, was not as effective as catheterization with air—which, as a bonus, was proven to be better tolerated by patients.

Itard and Deleau's work excited the profession and public, as newspapers printed reports of the "magic power" of catheteriza-

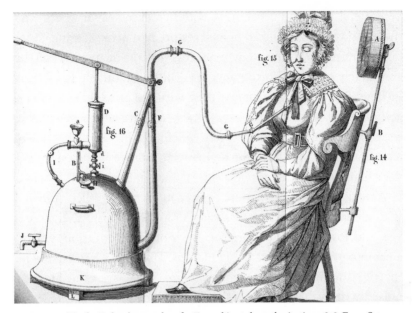

FIGURE 1.3. Nicolas Deleau's procedure for Eustachian tube catheterization, 1838. From *Sur le cathétérisme de la trompe d'Eustache, et sur les experiences de M. Itard, mémoire qui démontre l'utilité de l'air atmosphérique dans le traitement de diverses espèces de surdité* (Paris: Chez l'Auteur). Courtesy of Robert Ruben.

tion.[69] The procedure, however, was difficult for most surgeons to execute, especially in employing the catheter, which frequently induced vomiting and nausea in patients. Nevertheless, aural surgeons had another proven surgical procedure to offer their patients, and at the very least, Eustachian tube catheterization promised to be more effective than herbal-based emetics, nostrums, or other procedures such as syringing out the ear or perforation. But not all surgeons were eager to employ the technique. Some remained ambivalent about its use, citing a need for further studies to assess its risks and benefits. Others, like Turnbull, jumped on the bandwagon, professing its prowess as an absolute remedy for deafness.

Encouraged by Turnbull's enthusiasm for catheterization, patients routinely visited his office for several turns with the air pump. There were days when a crowd filled his office and waiting room, anxious for the miracle promised, scarcely thinking—or perhaps, not caring about—the dangers of the procedure. Sixty-eight-year-old William

Whitbread was tired of laboring with his "excessive deafness" and thus far catheterization had helped restore part of his hearing. Turnbull promised that after a few more sessions, the deafness would be completely cured. Unfortunately for Whitbread, on one fateful day in June 1839, something went wrong with the procedure. Immediately after catheterization, he was attacked with a violent swelling in the throat and died. A postmortem concluded that death was due to extensive inflammation of the brain owing to natural causes; a coroner's inquest found no wrongdoing on Turnbull's part.[70]

A week later, another patient expired in Turnbull's office. Afflicted with an ear irritation that was exacerbated by deafness, eighteen-year-old Joseph Hall had regular doses of catheterization with an air pump over the course of six months. During his last treatment, the air pump was filled four times before "there was a low gurgling noise" in Hall's throat, right before he "fell back, very gradually, with his head against the top of the chair . . . and did not speak, or move, or open his eyes again."[71] Witnesses were split on whether the fateful last pump was at the hands of Turnbull, an assistant surgeon, or Hall himself. Five prominent London surgeons performed the autopsy, finding evidence of a blood clot and apoplexy, but could not conclusively determine a cause of death.

It took a hundred years for Eustachian tube catheterization to be accepted by aural surgeons and less than twenty years for it to fall out of favor. Whether Hall's death was accidental or due to Turnbull's ineptitude, medical practitioners became increasingly aware that catheterization was no longer confined to any diagnostic or prescriptive guidelines. Gasconade and incompetence, combined with patient deaths, undermined its value as a safe treatment for hearing loss. But it was doubtful that the practitioners who routinely offered catheterization were even qualified to do so. In newspaper editorials, journal articles, and textbooks, aural surgeons either criticized or defended the procedure. Surgeon James Yearsley, for instance, insisted catheterization could have marvelous benefits for patients, but only if properly applied. In the hands of questionable men like Turnbull, only disastrous outcomes could be expected. As London-based aurist Alfred Barker affirmed, "The deaf have quite as much to fear from the inexperience of professional men, as they have from the more daring and less scrupulous tinkerings of arrant quacks and rougish knaves!"[72] Once practitioners ceased offering catheterization to patients, the procedure quickly fell out of favor among patients.

The Traveler's Potions

In an 1854 letter to his son Captain Dr. Edward Johnson, David Johnson of Limestone, South Carolina, confessed that his hearing was getting rather worse. Unsatisfied with the treatment options he had tried thus far, Johnson was advised by a reliable source that "there is now in Charleston a Physician (no quack but a scientific sense [sic] educated Gentleman) from London who is operating the ears of those who are born deaf."[73] The physician, Johnson wrote, had succeeded in enabling at least one of his patients to hear and speak; and he had "open[ed] the eyes of another who was blind." Undoubtedly, this physician is Alexander Turnbull, whose miraculous treatment offered eager patients opportunities for purchasing the latest scientific and medical advancements.

What Turnbull did after his exposure in Charleston, I don't know. He joins a long line of short-lived characters who gained notoriety before disappearing into obscurity. Medical mountebanks traversed the country, offering their pills and potions, stepping on stools and stages, promising healing and hearing. Their standard practice was to arrive in a new town, settle into some lodgings, and place an advertisement announcing their availability for consultation. Usually several testimonials were appended to the advertisement to give credence to this newfound cure, as Turnbull did with his "electro-stimulation" scheme. They were a thorn in the orthodox physician's side, taking fees from patients who were either ignorant of the professional discord or did not care. The ailing public already had a distaste for the physician—him, with his useless and expensive treatments, his bag full of purgatives and ancient herbal remedies. Far better to turn to the lay healer, to the health manual, or to the itin-erant passing through town. There was a better chance to cure deaf-ness.

Besides, there was no need to pay for a high-priced aurist and his fancy surgeries when an easy remedy could do the same job, but at half the cost and with minimal pain. There were plenty of "toadstool millionaires," as Oliver Wendell Holmes termed the proprietary manufacturers; these ended up on the "fringe" of medical practice, excluded from the medical mainstream for threatening the ortho-doxy of the "regulars," who were governed by legislation and profes-sional recognition. But the line dividing these groups was never fixed in place, especially since they selected from the same pharmacopeia to make their preparations.

Nor was the mountebank always a charlatan nor the charlatan always a quack. E. S. Aborn, a former surgeon at the Eye and Ear Infirmary of San Francisco, became a traveling physician in Penn-sylvania.[74] Newspaper notices announced his arrival at the Franklin Exchange or Merchant's Hotel, emphasizing his "world-wide reputa-tion" and his "guarantee of skills." Aborn, eager journalists asserted, had "performed cures in cases given up as hopeless years ago, by a single operation." But what made the physician trustworthy was his frankness in telling patients when a case was hopeless, to "neither put the patient up to such as expense, nor make him suffer the pangs of hope deferred."[75]

By the century's end, more miraculous remedies emerged, rooted in ancient texts but sprinkled with the sensation of modern science. The traveler's potion, like the surgeon's table, was attractive to deaf consumers seeking therapies for their usually incurable condition. Any chance of a cure, no matter how improbable, was preferable to a lifetime of isolation in silence. No wonder stories of new cures trans-fixed those seeking to alleviate their deafness, such as the deaf man who managed to sneeze himself into full possession of his hearing in 1872.[76] Not twenty years later, newspapers reported that Mr. E. H. Currier, the progressive principal of the New York Institution for the Deaf, was a believer in hypnotism as a cure for deafness. Although trance-like states for healing and religious practices date to ancient and probably prehistoric times, hypnosis as a healing form gained momentum after James Baird coined the term in 1841. Before the development of anesthetics, hypnotism was even used to put patients to sleep during surgical operations. For his female patients, the neurologist Jean-Martin Charcot even endorsed the practice as the best method for treating cases of hysteria.

Currier, however, saw hypnotism as an ingenious possibility for curing deafness. For if messages could be transmitted over wires, then wouldn't it be possible for the nerve to act as a messenger for the body? In a public statement, Currier stressed that "what we want, then, is knowledge as [sic] how these latest sensibilities can be developed the most rapidly and the most successfully. Now let us see what hypnotism will do."[77] He invited scientists and medical men working with hypnotism to experiment with pupils at the school. In the end, several experiments demonstrated that paralyzed auditory nerves, once properly developed through exercise, could strengthen enough to restore hearing.

Whether hypnotism was a remarkable cure or a fanciful wish, it was part of the trend of hucksters and self-styled healers who proliferated during the early twentieth century, depending upon mass appeals for funds through the mail. Faith healers joined this group. The city of Rochester during the 1920s was rife with stories of a certain Mrs. McPherson, who held several meetings in churches and houses across the city. Reports revealed she could draw on the power of the divine to cure deafness for all who came to visit her. "Little girls born deaf; youths deaf from infancy, some attending schools for the deaf—all cured instantly," a deaf instructor reported in a 1922 letter to the *Rochester Advocate* editor. "It is not a cure I am seeking," the instructor clarified, "but satisfaction concerning these stories. My parents believe them and think I should also be cured by miracles. I doubt them very much. Still miracles are not altogether improbable."[78] After making inquiries and examining pupils at the Rochester School for the Deaf whose parents or family members took them to visit Mrs. McPherson, the editor replied to the instructor that "we have to report that we could not find the hearing of any child had benefited in the slightest degree."

For the religiously devout, sometimes turning to the divine was the first option, if not the only one, for dealing with bodily ailments unresponsive to surgical or medical recourses. Religion occupied a central role in my family, directing daily life with regular prayers and weekends in Punjabi school. We recited the *ik onkar* prayer, the first prayer of our holy scripture, to show piousness and received the repeated message that fears of the unknown and all of life's worries or expectations, can be kept at bay by uttering the phrase *satnam waheguru ji* several times. *Waheguru*, my mother would always tell us, is bigger than our worries.

2

Ear Spectacles

If the wish to ignore deafness was not so pronounced, ear trumpets would be as common as spectacles.

Michigan Farmer, 1882[1]

Conspicuous and annoying as trumpets are, there is a great search for invisible yet potent helps; and the hundreds of thousands spent in advertising such devices give strong testimony to the far greater sums that must constantly pass into the hands of those who offer new patented helps.

B. Alex Randall, 1905[2]

Harry Stephen Keeler's eccentric 1939 novel, *The Man with the Magic Eardrums*, unfolds with wealthy bookmaker Mortimer Q. King returning to his Minneapolis home under cover of night to find a burglar entering through a window. While holding the burglar at gunpoint, King caught sight of "a peculiar little device of some sort stuck into his ear." He observed it was made of hard rubber in a cone or cylinder shape and was "ingeniously flesh-colored—and therefore as good as invisible against the actual flesh of his ear."

"What are those things—in your ears?" King asked. The burglar gave the bookmaker a pained look, not wanting to answer, but nevertheless replied: "I don't mind telling you, Mr. King, that if you weren't sitting there with a gat near your mitt, I'd tell you you were nuts as hell—and that there wasn't anything in my ears—and that if there was, it wasn't any of your goddamned business! However—you're holding aces right now. So all right. They're—they're Cromley

Patented Eardrums—at least I call 'em drums—they're really little trumpets—for deaf people—people, that is, with certain kinds of deafness."[3]

The burglar revealed that he was completely stone-deaf such that if he removed the drums, he could not even hear a "police siren blown ten feet" from his face. Yet the drums afforded him a secret power: he could hear sounds otherwise inaudible to others, sounds like the creaking of the electric mechanisms caused by magnetic stresses in the five-dial burglar-proof safe kept in the King residence. Upon hearing the burglar's story, King offered him a deal: if he opened the safe, King would allow him to escape instead of calling the police. After all, King was in the house with the same intention: to steal his wife's jewels, which were kept in the safe.

Keeler's work of intrigue, with its highly complex, bizarre, and nearly inscrutable plot, is pulp fiction, but the deafness of burglar Peter Givney certainly must have resonated with deaf readers. Givney's story line is a twist on standard portrayals of deaf characters and their struggles to navigate the hearing world: with his magic eardrums, Givney is privy to a world of sounds unavailable to ordinary ears. Using acoustic devices, denying the impairment or even making use of clever techniques such as lipreading may disguise one's deafness, but at the same time, they stigmatize it. Revealing the hearing loss would reveal one's secret self. For Givney, on the other hand, his devices do more than lend him hearing—they make him superhuman. For the moments he wore the magic eardrums, he could even consider himself normal.

Though fictional, Cromley Patented Eardrums were indistinguishable from similar products appearing in advertising sections of local and national periodicals during the first half of the twentieth century. From the early eighteenth-century appearance of proprietary brands, to the surge of packaged nostrums after the Civil War, patent medicines maintained a significant hold on American culture and health. "Patent medicine" did not always mean a formula was patented. Patenting a nostrum meant a proprietor was required to reveal ingredients for transparency. Since secrecy provided a competitive advantage, sellers frequently took to stamping their product as patented to protect their proprietary makeup. Most patent medicines were shipped from Britain, arriving in the colonies on ships and stocked by apothecaries, postmasters, grocers, and physicians, and advertised in listings of patent medicine brands.[4] Trade disruption during the American Revolution ended reliance on English medicine and marked the emergence of cheap American imitations—some of which were retailing long before the disruption but were then rebranded and concocted for wholesale or retail druggists as "new" remedies. By 1790, the U.S. passed a Patent Act (revised in 1836), creating the modern patent system that acknowledged the commercial potential of an original inventor's designs by providing them monopoly protection against competition.[5]

The commodification of medical goods allowed deaf consumers to self-prescribe and self-medicate. Advertised cure-alls, specifics, and nostrums provided them opportunities for discreet relief. Where those failed, a host of sound magnifiers, auricles, invisible acoustics, and every kind of trumpet as relief for "every degree, for church, general conversation, &c.," were available as prosthetic extensions for damaged ears.[6] With the merging of medicine and modern industrial capitalism after the 1860s, the design, manufacturing, and sales of prosthetics and assistive devices flourished on both sides of the Atlantic.[7] A variety of prostheses (some of which were patented by their makers), historian David Turner explains, "were increasingly advertised to the public in ways that emphasized their value in restoring function and the appearance of normality," while also addressing the polite aspirations of users.[8]

Aural prosthetics—or, "aids to hearing," as advertised—were specially designed to be conspicuous. Around 1800, German manufacturer Frederick Charles Rein set up shop in London as an "acoustic inventor," selling items of various sizes and shapes, including long trumpets sixteen to nineteen inches in length, pocket collapsible trumpets, and banjo-style or bell/dome styles that were more powerful for resonating sound. Later renamed F. Charles Rein & Son and marketed as "the originator of scientific deaf aids,"

FIGURE 2.1. Advertisement for F. C. Rein, one of the largest manufacturers of acoustic aids in England. CID–Max A. Goldstein Hearing Collection VC 703, box 48, folder 703426. Courtesy of Bernard Becker Medical Library, Washington University School of Medicine.

for nearly a century the company was the leading manufacturer and seller of hearing trumpets in England and the United States before closing shop in 1963. Their research unit designed and perfected models of acoustic aids that were the epitome of modernity, lending scope to craftsmanship by offering individualized fittings rather than mass-produced imitations. The company also offered acoustic aids that were designed to be hidden—concealed within the skin or body, or disguised as furniture and ordinary objects, crafted to "disappear" when not in use.[9] Urns, fans, thrones, headbands, and walking canes contained cleverly obscured ear trumpets. Some were remarkably beautiful, though more ornamental than functional, such as the company's signature aurolese phone floral tubes designed to be worn within a woman's bouffant hairdo or underneath a hat. Despite its fragility and small size, this device provided an acoustic gain of nearly ten decibels, a slight improvement for female users with mild hearing loss.

Rein's products were, however, pricey and beyond the reach of most working-class consumers. Those looking for inexpensive alternatives sought out Thomas Hawksley's shop in London, or George P. Philips & Sons of Philadelphia. Yet no matter the size of device or its price, acoustic aids required the user to devote time and effort to ensure that conversation could become intelligible without the awkwardness of lipreading or writing out the conversation.[10] Most users who lost their hearing gradually or in old age, however, found these acoustic aids problematic: they were bulky and cumbersome, serving as both objects of pity and reminders of infirmity. At times, the aids could be painful to use and even forged unhappy relationships between individuals and their prosthetics by creating a lasting sense of shame and self-stigmatization. Yet it was still better—wiser, even—as the Globe Optical Company of Boston asserted in 1890, "for a deaf person to get some instrument, to *aid* the hearing, rather than to draw attention to the infirmity by the necessity of loud talking."[11] Revealing one's deafness in polite society was worse than wearing a prosthetic aid to hearing.

While most aids to hearing were designed to be camouflaged, artificial eardrums were crafted to rehabilitate an invisible impairment without drawing visible attention to the user. They initially arrived on the market during the 1850s as a newfound medical treatment for cases of deafness caused by eardrum perforation, but by the century's end, they were reshaped and recommodified as prosthetic products of the new "scientific age." Sold alongside trumpets, auricles, and domes, artificial eardrums were marketed as a triumph over mechanical relics. They were, in other words, "scientific sound conductors" that discreetly cured deafness at a fraction of the cost of medical treatment. Sometimes they were pretenders, these hearing devices that mimicked the organic drum, pressuring their position and causing further damage despite their insistence of prosthetic assistance. They straddled the vexed line between prosthetic and cure: as extensions of the ear, they were "inserts" or "invisibles," even "machines," "ear phones" or "tiny megaphones," but once worn, they (temporarily) eradicated the affliction of deafness.

As with other acoustic prosthetics, artificial eardrums allowed users to conceal their deafness, both figuratively by enabling them to "pass" as hearing, and literally, by correcting the defective ear-drum.[12] Thus, they were marketed as ear spectacles with the potential for delaying the "anxieties of deafness"—an item of relative ease, but also of acceptance, as eyeglasses are not traditionally perceived as disabling. The medical publisher J. M. Churchill elaborated in 1853: "When we reflect on the economy of the eye and on the benefits derived from the mere use of glasses, does it not appear possible that dullness of hearing, and some cases of deafness may be relieved by the employment of an acoustic apparatus?"[13] Acoustic aids and artificial eardrums—and later, hearing aids—cannot restore perfect hearing the way glasses can restore sight, but promoting them as ear spectacles serves to normalize the object, make it more acceptable, more visible, not just to others, but also to users who denied their deafness.

"It has always seemed to us strange," says the 1900 catalog of instrument makers William V. Willis & Company, "that people object to letting their friends know they are hard of hearing, when the use of these simple devices would bring so much happiness to themselves."[14] If people wear spectacles without any feeling of difference, then why shouldn't those needing aids to hearing also experience the same, rather than remaining irritable and unhappy because of unsatisfactory communication? Why, as historians Graeme Gooday and Karen Sayer ask, don't we refer to the blind or partially sighted as "hard of seeing"? Perhaps the difference resides in the labor, for "the use of hearing aids has generally required far more skill, patience, and effort than the successful wearing of glasses."[15] Furthermore, hearing loss tends to be more stigmatized than vision, due to the close associations between dumbness and deafness.

Then there were some users who wholeheartedly embraced their acoustic prosthetics as inseparable, ingrained extensions of their body, even their identity. They framed the devices as badges of pride, sometimes crafting delicately woven and ornately beaded purses for storing their prosthetics when not in use, as Amelia Woods, a Canadian deaf woman, did in the 1880s.[16] As extensions of the ear, hearing prosthetics certainly can connect deaf people to the world of sounds, but they also create a dependence: a user is "cured" or has "normal" hearing only when the aid is in use. When it is not—when it is placed on a lap, put away in a drawer, or forgotten somewhere— the user is once again deaf, drowned in silence.

Sometimes the silence is of your own choosing, crafted at guarded moments, when the aid comes off and an exhausted sigh escapes your lips; you crave the silence, crave the freedom of your ears without a plastic object in it, a break from a cacophonous world.

Brett and Toby

The dependence of some users on their acoustic aids enabled them to craft new identities that allowed them to control, if not define, the discourse surrounding their deafness. These users were not passive recipients of a medical device, but rather, active participants who evaluated, modified, and in some instances, even designed their own prosthetics. Prosthetics then, became culturally symbolic, part of the user's quest for "functional normalcy," that enabled them to assert their autonomy, personhood, identity, and self-presentation— even if it was just to "pass" as hearing.[17]

Dorothy Eugénie Brett, the daughter of Reginald Baliol Brett, 2nd Viscount Esher (Lord Esher after 1899), close adviser to Queen Victoria, and Eleanor van de Weyer, spent the first twenty years of her life cloistered under the powerful influence of her father. Born on 10 November 1883 in fashionable Mayfair, London, Dorothy was from the beginning a proud, haughty, and bad-tempered child. Her doll-like features—golden hair, luminous brown eyes, and pouty lips—scarcely subdued her rebellious nature. She was affectionally called Doll by her family and spent her childhood days playing in the corridors of Windsor Castle with regular excursions to the Red Drawing Room for dancing lessons with her sister and Princess Beatrice's children; old Queen Victoria frequently made an appearance, watching them and tapping her cane in unison with the music. Though fine with the children running around, the queen frowned upon the misbehaving, especially when Dorothy "knocked little Prince Leopold over," prompting the "anxiety of the nurses as they picked him up; he bled so easily," the result of hereditary hemophilia.[18]

On 22 January 1901, Queen Victoria died at the age of eighty-one. A year after the funeral, during preparations for the coronation of Edward VII, Dorothy fell ill, "stricken with violent pains in [her] tummy." In her unpublished memoirs, she recalls this moment, when her father was too busy with the coronation arrangements and her mother too aloof to notice her sickness:

> I am so ill that I am in bed, I think, for two weeks with this tremendous pain. I can remember the agony of it. I can remember almost trying to crawl up the wall when the pain swept back on me. It would come and go in great sweeping movements. A nurse is procured and brought from Windsor. She is a very nice woman, a very intelligent woman, an excellent nurse. She is horrified at the sight of me, my tummy all swollen up and I in such pain that there was nothing she can do. Finally she goes downstairs to my parents. She tells them that if nothing more is going to be done she is leaving, because she does not think that I am going to live and she cannot be responsible for me.[19]

Frightened by the nurse's announcement, Esher cancelled his meetings with the king and rushed back to the family home in Orchard Lea. Edward later sent a telegram expressing his sympathy over Dorothy's health and advising Esher to call for the king's personal physician from London, Sir Frederick Treves.[20] One of the most prominent physicians of his time, Treves initially made his name as an anatomist and for his care of Joseph Merrick, a man with severe deformities who was billed in the media as the "Elephant Man." His position as Royal Surgeon further solidified his reputation; fees received from private patients were sufficient enough for him to resign from his position at the London Hospital.

Upon examining Dorothy, Treves diagnosed her with a severe case of appendicitis and declared that she needed to be operated

on immediately. Despite his expertise, Treves knew surgical removal of the appendix was risky with only a 10 percent chance of
success. Delaying surgical intervention was just as perilous, a lesson
Treves tragically learned in 1900 when his eighteen-year-old daughter Hetty died following an appendectomy that was performed too
late. Treves urged Esher to approve the procedure.[21] "Another room
is scrubbed out, as there is no time to take me to a hospital," Dorothy
recalls in her memoirs. "The room is cleaned, a table put into it."
After she received an anesthetic, Treves proceeded: upon opening
her, he realized Dorothy's appendix had already burst before he
was called for. Following the operation, in the "most awful mess,"
Dorothy spent several weeks recovering with a large tube in her side
that gradually drained out fluid. Confined to a wheelchair during her
long convalescence, she nearly missed the king's coronation ceremony—but then the king fell ill, and it was feared it was appendicitis. "At the time it was thought that perhaps my father had carried a
germ from me to the King, so little was known of that disease at the
time," she recalled. Two days before the king's coronation, Treves
performed the operation despite great pressure to postpone until
after the coronation.

Although Treves's surgery saved her life, Dorothy blamed appendicitis for her gradual hearing loss, which she believed first began
with childhood trauma. "I think my deafness began with the desire
to shut out the criticism and bullying of my brothers," she wrote.
"Purely psychological. . . . I simply desired to shut it all out." Deciding to shed her secluded upbringing, she enrolled in the Slade
School of Art in 1910. She cut her hair in the Bohemian style peculiar to Slade students, one that led Virginia Woolf to label them all
as "cropheads." And in the tradition of the school, which called for
students and staff to be known by their surnames, Dorothy became
Brett, embracing a new identity to distance herself from her aris-
tocratic privilege. She took to wearing men's trousers, a rebellious
move for a woman of her stature and time. Brett counted members
of the Bloomsbury Group and other artists and writers among her
friends: Virginia Woolf, George Bernard Shaw, Lady Ottoline Mor-

rell, Bertrand Russell, Aldous Huxley, Katherine Mansfield, John Middleton Murry, Dora Carrington, Mark Gertler, and D. H. Lawrence.

At Slade, Brett's hearing declined significantly, though most of her friends did not recall her using an ear trumpet or aid of any kind. Indeed, it's likely Brett perceived her hearing loss as a temporary affliction, one that could be cured with proper medical care, as she told her father that she planned to see an aurist to try "a radium cure—according to the *Daily Mail* marvellous cures have been effected."[22] She even tried Doctor Foster's Miracle Cure, composed of "injections in my behind," and a trip to the tropical airs of Sarawak with her sister Sylvia, hoping to "return with my deafness cured . . . because really I don't see much use in being alive and stone deaf." The trip, however, was cancelled after Sylvia fell ill.

By 1915, Brett began to frequently withdraw into her deafness: "It was at this time that my deafness grew worse and I felt that I was slowly being cut out of life." She attributed the change to her awakening, the realization that her deafness was a manifestation to protect herself against her brothers' emotional and physical bullying. She found herself hopelessly trapped between two worlds, one of inner security, the other a longing to be involved with the vibrant and intellectual spirit of her friends. "She was of the squirrel type, and poor thing, very deaf," Lady Ottoline recalled years after her friendship with Brett waned. "I liked her and felt sorry for her, as her deafness made it difficult for her to be friends with people, and I knew her family were not very kind or sympathetic to her. I felt she was a sort of Cinderella, and that I had to help her find new friends."[23] Not everyone was sympathetic to Brett's deafness. Bertrand Russell, for instance, would get angry if she left to retreat into her deafness at the precise moment he was going to read from a book, urging her to return and sit close to him to hear every word. When Russell was in prison before the end of the Great War, Brett wrote him a long letter detailing her experience with deafness, comparing her hearing impairment with being in prison:

But you can imagine what it means to see life revolving round you—
see people talking, quite *meaninglessly*! Like looking through a shop
window or a restaurant window. It is all so hideous I sometimes wonder how I can go on. I think if it were not for my paintings I would
end it all.[24]

Russell's reply was only half-sympathetic. He urged her to practice lipreading, take care of her "inner attitude," and prevent herself
from being aloof or isolated. "Though your deafness may make that
harder, it doesn't make it impossible."[25] Perhaps he, like others in
their circle, perceived Brett's deafness as a case of psychosomatic
selective hearing, or that she "imposed deafness" whenever she met
an intellectual. Brett acknowledges this possibility in her memoirs:
"It may very well be true. . . . My deafness began at the time I did
begin to meet these engagingly frightening intellectuals who made
up the world of painting and literature."[26] Her personal accounts,
however, reveal a more depressive state, a woman caught between
two worlds and striving to normalize herself through her hearing.

Those who knew Brett also knew Toby, the flat tin ear trumpet
she carried with her to aid in conversations. That she named her
trumpet and referred to it by name in her correspondence indicates
the intimate connection she cultivated with her acoustic aid, perhaps as an escape from her inner sanctuary. She experimented with
other acoustic devices, including a Marconi listening machine, but
her friends remembered her reliance on Toby—and their shouting
into it—more so than her use of any other "machines." While visiting
Ravello with D. H. Lawrence, Toby was with her as they drove by a
site known for a famous echo, though Toby could hardly assist Brett
in hearing the noises. Toby made an appearance in her self-portrait:
wearing a dark cloche hat and a raincoat with the collar drawn high,
Brett stands tall and fixes her gaze on her audience. One hand rests
inside a jacket pocket, the other grips Toby. In a 1921 photograph
supposedly taken in Sierre, Switzerland, writer Katherine Mansfield
is seated on a wicker chair next to Brett, who is resting Toby on her

FIGURE 2.2. Katherine Mansfield with Dorothy Brett, who has Toby in her lap, 1921. Ida Baker Collection, Alexander Turnbull Library (Ref: 1/2-011927-F). The file records indicate that this photo was taken in Sierre; however, the library acknowledges that there is no mention in any of Katherine Mansfield biographies that Brett was ever in Sierre in 1921.

lap. The two women swore an eternal pact of friendship and were so close that Mansfield empathized with Brett's deafness, writing in a letter "how deeply I feel for you in your deafness." "Perhaps you think people 'accept' it, forget it," she continued. "I never do and I never could. I think you are wonderfully courageous to accept it as you do and I am constantly realising what it must mean to you."[27] In a later painting, *My Three Fates* (1958), Brett depicts Mabel Luhan, Frieda Lawrence, and herself in a room, with D. H. Lawrence seated underneath a great pine tree at Kiowa Ranch; gray-haired Brett works at a typewriter (she often typed Lawrence's handwritten manuscripts), and Toby is long gone, replaced by the improved electric hearing aids, one of which is visible in Brett's ear.

Deafness daily "brought on an excessive tiredness after so long watching people's mouths continually opening and shutting."[28] Later in her life, she penned a letter titled "Christmas Morning," acknowledging her dependence on Toby and later, her electrical hearing aids: "I am tiresome to my friends but mostly to those who yell at me, until my instrument [vibrates] and buzzes like a bee-hive. It

FIGURE 2.3. *Self-Portrait* by Dorothy Brett, 1925. In private collection, courtesy of Addison Rowe Gallery.

FIGURE 2.4. Dorothy Brett with one of her "machines": a radio that she could attach her hearing aid to. Overjoyed about being able to hear clearly, she spent days listening to different programs and writing about her experiences in letters to friends. Mable Dodge Luhan Papers, YCAL MSS 19, Yale Collection of American Literature, Beinecke Library.

is a natural, instinctive thing to shout at the deaf, but when tuned in I am no longer deaf."[29] Her hearing aids were central to her life, and at times she even placed them on the center of a table to ensure that they dominated conversations between friends, as she did with her Marconi machine and a 1940s Sonotone hearing aid she kept in

FIGURE 2.5. Dorothy Brett in Taos, New Mexico, c. 1957. She is wearing an electric hearing aid and standing next to her painting, *My Three Fates*, which depicts herself wearing an older hearing aid model. Dorothy Brett Collection 1898–1968, box 2, folder 5, Harry Ransom Center, the University of Texas at Austin.

a carved leather case decorated with pieces of Indian silver and turquoise. Brett's biographer Sean Hignett, upon meeting the painter in 1970, observed that she was still using a hearing aid resembling a scaled-down 1930s radio that she hung around her neck.[30]

Toby, however, occupied a special place in Brett's life. Indeed, she was so completely beholden to Toby that the loss of the device devastated her. In 1924, she accepted Lawrence's invitation to join his utopian community Rananim, located at a ranch at the Sangre de Cristo Mountains north of Taos, New Mexico. Fourteen years later, she would become an American citizen. While touring Mexico with Lawrence and his wife Frieda, Brett was dismayed at Lawrence's comments that perhaps "the Virgin of Guadalupe will cure your deafness." She didn't of course, but Brett noted that the fiesta was lovely around the Old Church, and that the "many paintings of miracles absorb us." At the crowded station of Oaxaca, while

heading towards the Hotel Francia, much to her dismay, Brett realized "Toby is missing—my precious ear trumpet. I rush back to my
room; I search everywhere. . . . Not a sign. I am in despair."[31] "I was
helpless," she recalled. "Lawrence was furious, such things upset him
and made him angry."[32] (While Lawrence seemed to accept Toby, he
stigmatized Brett's use of mechanical aids, stating, "I cannot think
of anything to say to a black box.") The managers of Hotel Francia
plastered the town with notices of Toby's theft but told Brett that it
was unlikely that Toby would be found and returned. While waiting for a replacement to arrive from London, Toby was cloned by an
"intelligent Indian" whom Brett commissioned to make the device
from her sketch.

Upon returning to her New Mexico ranch, Brett described the
"strange communication" she developed while listening to the
silence around her, through Toby, and without: "The silence around
me was not the silence of deafness, in spite of my not hearing the
flow of the water, the birds, and the animal calls. One becomes aware
of vibrations—of a different, more subtle form of communication."[33]
Brett's personification of her ear trumpet certainly speaks to her
polite aspirations to maintain an appearance of normalcy within the
confinements of her social groups. Through Toby, she used a range
of skills—lipreading and aided listening—to effectively mediate her
lived experiences with deafness. In so doing, Brett could control
the discourse surrounding her deafness, thereby positioning Toby
in her own quest for functional normalcy to assert her autonomy,
personhood, and self-representation, even if it was merely to "pass."
As Brett grew older and distanced herself from her Bloomsbury
friends, she became more confident in her identity as an artist, eventually embracing the newest electronic technologies. She no longer
struggled to conceal her devices, nor did she rid herself of the Toby
clone, continuing to rely on him even as she wore hearing aids.[34]

From the upper echelon of Victorian aristocracy to the rugged
landscape of Taos, Dorothy Brett referred to her personal odyssey
as "a long and beautiful" journey.[35] Toby accompanied her across the

sea, safely tucked away in her hand, or lap, until his disappearance.
During their time together, the two were inseparable.

Ear Extensions

Generations before Brett and Toby, Scottish geologist James Hutton
sat for his portrait for the Moravian Society with his trumpet applied
to his right ear.[36] Hearing impaired most of his adult life, eventually
Hutton's hearing deteriorated to the point that he could converse
only with the assistance of an ear trumpet.[37] Like Hutton, there are
others who brought their trumpets with them for their portrait sit-
tings. Eighty-eight-year-old W. P. Zuber, for instance, was photographed in 1900, sporting a white beard and holding an ear trumpet
to his ear.[38] There are more nineteenth-century images circulating
around the internet (though it can be difficult to confirm their au-
thenticity): a middle-aged woman in a bonnet seated and clutching
her tin horn; a carte-de-visite of an elderly woman holding her conversation tube; an albumin print of a woman seated at a desk, her
ear trumpet in right hand; a family gathering outside, with grandma
speaking into grandpa's conversation tube; and a portrait of a senior
couple seated in a garden, the man holding a large ear trumpet in his
lap.[39] While these images speak to different experiences of hearing
through technology and tell us how deafness was not always a condition to be ashamed of, hidden away, or concealed, it is also rare to
find images of younger individuals with their acoustic aids, another
indication of course, of the association of hearing loss and old age.

The first time I tried a historic ear trumpet, I was surprised that
it was a lot lighter than I imagined it would be. Resembling a banjo,
this early twentieth-century tin version was far from the beautifully
decorated brass versions I was accustomed to seeing in museum collections. It was crudely constructed, with some of its metallic parts
poorly melded together. I traced the edges of the trumpets, thinking
about how similar it looked to Brett's Toby: a flat disc on one end,
capped with another half-disc for collecting sounds and channeling

FIGURE 2.6. Photograph of an elderly woman holding her conversation tube in hand, c. 1890–1910. Courtesy of Phisick Medical Antiques (www.phisick.com).

them through the funnel to the rubber earpiece. Generally, ear trumpets operated on a simple principle for amplifying sounds: the larger the receiver, the more sounds were channeled, and thus, the louder they became. As I held this tin trumpet in my gloved hands, I wondered how much its former owner relied on it, how much she cared for it, how dependent she was on it. Since this model could be dismantled for portability, I imagined its owner taking it out from a bag or drawer only when she needed it, perhaps only in conversational settings where she felt she would not be judged.

The curator initially hesitated to allow me to try out the trumpet, no doubt with liability concerns in mind. I was reminded of a 2003 study at the Central Institute for the Deaf at Washington University in St. Louis, where archivists and curators tested out the decibel ranges of various nineteenth-century ear trumpets and acoustic devices, measuring their effectiveness by how well sounds were amplified.[40] Their assessments were insightful for analyzing how helpful acoustic aids were for people with mild to profound hearing loss. By testing the acoustic gain of the devices, we could also learn how construction styles—curves, sizes, funnel shapes—behaved for transmitting sounds. The tin trumpet I held in my hands, which I learned was sold as "Miss Greene's Hearing Horn," was presumably an effective option for hearing-impaired people, as the bulb-shaped discs could receive a wide range of sounds. The trumpet was inexpensive too: the large version was advertised in the 1902 Sears Roebuck catalog for $1.32—about $40 in 2018 terms.[41]

"I'll be really gentle," I pleaded with the curator. "I only want to know if I can hear with it." After what seemed to be the world's longest pause, the curator reluctantly agreed, standing close to my desk to ensure that no damage would befall the trumpet. Excited, I called my colleague Kira over to assist me. "I'm going to take my hearing aids out," I explained, "and try out the trumpet. Could you speak to me in a regular tone? Let's see if I can hear with this." Kira chuckled at my eagerness and agreed to participate. I removed my hearing aids, carefully picked up the trumpet and placed the earpiece in front of my ear canal—not inside, because of the trumpet's fragility and hygiene concerns, even though I knew this would distort the experiment. I strained to listen, my back turned from Kira so as not to lipread in my silence. After a few seconds, I turned around; Kira had spoken but I had heard nothing. I'm severely deaf, so perhaps an ear trumpet was not for me. Raising the trumpet upwards, as I've seen in historical photographs, I asked Kira to speak directly into the trumpet.

Sounds! Faint, but still! I made out a "hello, Jai," and some garbled speech. We then tried out different positions with the trumpet—slightly bent towards the ground, high and angled away from my ear. Later, as I learned more about Miss Greene's Hearing Horn, I noticed that the advertised photographs of a woman using the trumpet had her holding it low, close to her collarbone and chest, a position that for me delivered no audible sounds. She seemed bored, her expression a far cry from the overjoyed one I sported in the photograph Kira took of me. I wondered if the woman in the advertisements was deaf and if she was, whether she even heard through the trumpet when it was positioned that way. Perhaps not. However, holding it upwards to listen, as I did, could certainly be tiring on the arms for prolonged periods.

The uncomfortable placement of the ear trumpet and the poor acoustic gain of some designs would certainly have created awkward situations between listeners and speakers that complicated social boundaries of propriety and individual space. The design trend favoring camouflage and invisibility in acoustic aids would have been appealing to users wanting to limit the embarrassment of such situations or revealing their deafness—especially women who were expected to maintain appearances suitable for their social position. Trumpets masked with lace and ruffles, for instance, were designed to be hidden in a woman's dress; black lace would enable the trumpet to become an elaborate accoutrement for mourning during the Victorian period. Later in the twentieth century, hearing aid firms sent out free pamphlets advising women on how to adopt hairstyles of "fashionable flattery" so they could disguise their aids, and digital hearing aids were made in various colors to be concealed against a range of skin tones or hair color.

No matter how well a deaf person carefully crafted social situations generated by their deafness, their ear trumpets were obvious markers of difference that aroused pity in their hearing counterparts—as they were for Edith Ella Baldwin, an American art student who resided in the Academie Julian in Paris with her sister Katherine during the 1890s. A chance encounter introduced Baldwin to deaf watercolorist Elizabeth Eleanor Greatorex, daughter of the distinguished American painter Eliza Pratt Greatorex. In her unpublished memoirs, Baldwin recalls the encounter: "Well the Miss

Greatorex I met is deaf and almost dumb; but she is very bright and clever, and it is pitiful to see her struggling to say something she cannot. She uses an ear trumpet, and I feel sorry for her."[42] In another diary entry, Baldwin recalls encountering Greatorex at a gallery showing: "We met a great many of our fellow students there and Miss Gretorex [*sic*] also. She is deaf and uses an ear trumpet. We like her very much but we are a little afraid of an ear trumpet, and our endeavors to make her understand through that are amusing to say the least." One doubts it was amusing for Miss Greatorex.

When artificial eardrums appeared on advertising sections, they posed an attractive alternative to awkward and bulky acoustic aids. For one thing, they were cheaper than their counterparts: they ranged in cost from $3 to $5 whereas trumpets were listed in trade catalogues for between $8 to $25 or more, depending on the make and model. They required minimal adjustment, usually at insertion in the ear canal, to guarantee maximum sound resonance. Users testified that despite the initial overwhelming abundance of sound, they could hear noises they had never experienced before or had heard only faintly. Some described a dizzying sensation that increased sensitivity to both sounds and the presence of a foreign object in the ear. Even noises once heard only in the background were amplified; middle-class women, for instance, remarked hearing the "intolerable rustling" of their silk dresses—noises that young male patients found rather astonishing.[43] A housekeeper described being able to hear cacophonous street noise, including footsteps on the pavement and the words of people who passed by her. For some users, artificial eardrums completely normalized their hearing, effectively "curing" them when worn, while for others it merely amplified sounds as much as twenty to twenty-five decibels.[44]

Artificial eardrums, however, were not initially designed as a commercial cure, but rather as a surgical prosthetic to be carefully regulated and managed by aural specialists for specific cases of deafness. First advertised in 1848 as an "artificial tympanum" for deafness caused by eardrum perforation—a common ear injury that usually heals on its own, the result of infection, trauma, or shock of loud

noises—the device originates in folk remedies advising the insertion of organic materials into the ear to serve as prosthetics for eardrum rupture. Healers recommended fish skin, pig's bladder, egg
membranes, wetted paper, or onion sprouts. In 1640, German physician Marcus Banzer developed a tube constructed out of elk hoof
wrapped in a pig's bladder, a prosthetic that was occasionally cited
by physicians in the preceding centuries, though not necessarily prescribed. None of these remedies promised a permanent fix for perforation, but the idea of a prosthetic was appealing. The challenge
was to find the right sorts of materials to create a device strong and
durable enough to replicate the natural flexibility of the eardrum.

In 1848, British aural surgeon James Yearsley published an article
announcing his creation of an artificial eardrum after being inspired
by a patient's complaint about hearing loss following an eardrum
perforation. After experimenting with different materials, Yearsley
constructed a cotton wool pellet that, when soaked in glycerin and
inserted into the ear, could shape and harden against the perforation, thereby restoring hearing. For this patient, and the many others
who visited Yearsley, the prosthetic functioned as a cure. Indeed, the
patient's children, perhaps accustomed to their father being unable
to hear properly, were surprised at the strength of the new prosthetic: "Your ears are too sharp; we cannot speak to mamma, even a
whisper."[45] In 1852, four years after Yearsley published his findings,
surgeon-anatomist Joseph Toynbee introduced his "artificial membrana tympani," a device emerging out of extensive pathological research on diseased ears: crafted of two fine silver pieces affixed to a
wire with a layer of vulcanized rubber between them, Toynbee's design was more durable and flexible than Yearsley's. Though the two
men—who happened to be neighbors living on the same street in
London—publicly squabbled over priority, Toynbee's design would
eventually become the forerunner of all commercialized artificial
eardrums that entered the market by the end of the nineteenth century.[46]

It took time for ear specialists to embrace the benefits of artificial eardrums, primarily because both Yearsley and Toynbee insisted

that the device was applicable only to cases of deafness caused by eardrum perforation. It was important for the device not to be marketed as another catch-all cure for deafness. Not only would it be ineffective in other aural cases, but patients risked further damage if the device got stuck in their ear canal. While Toynbee's esteemed status as a surgeon provided credibility for the commercialization of artificial eardrums—if only within medical circles *and* only for perforation—other surgeons inflated the benefits of the device. Glaswegian surgeon Thomas Barr, for instance, claimed that except for extreme cases of permanent deafness, the artificial eardrum was useful for enabling deaf persons to hear ordinary conversations perfectly well, perhaps better and more comfortably than with an ear trumpet.[47] Furthermore, since the ethics of medical commercialization often conflicted with the challenges that quackery presented to the credibility of cures, practitioners suggested that if it was important for artificial eardrums to be regulated, then a more general design was required, one that could benefit cases of chronic middle-ear disease resulting from damage to the ossicles. British surgeon John Ward Cousins thus designed his "Antiseptic Artificial Eardrum" as a safe alternative for frustrated patients leaning towards selecting questionable artificial eardrums listed in the advertising sections of periodicals. After all, if patients were going to forgo expert advice, then the least aural specialists could do was provide an affordable and verified option for them.[48]

Thirty years after Yearsley and Toynbee introduced the artificial eardrum to their medical brethren, aural specialists and their associated British and American manufacturers struggled to regulate sales of the device to prevent its marketing as a catch-all cure. Manufacturer Thomas Hawksley published a pamphlet with a disclaimer that artificial eardrums should be purchased only upon consultation with a skilled aurist.[49] Surgeon John Nottingham, however, added that despite aurists' intentions, sometimes patient expectations could exceed the "curability" of a device, for the "excitement, or hope" at first application "may lead to the expression that the hearing is improved, but which subsequent observations will not always confirm," especially when patients were disappointed to discover their deafness "returned" when the artificial eardrums were removed.[50]

Hope for the Deaf

At the corner of St. James and Belgravia Courts in Louisville, Kentucky, stands an eclectic brick mansion with Chateauesque details, including a spiked and steeply rounded turret roof, tall windows, and a foyer opening with a grand staircase. Built in 1891 as an exclusive club and casino for gentlemen residents of the area, the property was eventually sold, changing hands until one of Louisville's more colorful residents moved in: George H. Wilson, president of the Louisville School Board, vice president of the Roberts Brothers Oil Company, and one of the most successful businessmen in the city.[51] The Wilson family resided in the house in the early 1900s, upgrading the property (including spending $8,500 for a new backyard with a pool) and adding a brick wall and glass in the turret.[52] George Wilson secured his status as a prominent businessman by becoming the first person in his neighborhood to own an automobile. The house was later sold in the 1920s to the Women's Christian Temperance Union; they painted the building and bestowed it with a nickname—the Pink Palace.

Within his business circles, Wilson was considered a shrewd entrepreneur and inventor. In 1892, he obtained a patent for his "rimless and self-ventilating" artificial eardrums and promoted them in the daily newspapers as "wireless phones for the ears." Made of rubber and designed to be invisible and nonirritating when placed in the ear, the device was marketed as one of the best "sound conductors" for the ears. The company's early advertisements featured a cross-section of the ear with the device in place to maintain the rhetorical gauge between visibility and invisibility: a customer could see the benefits of the product and wear it comfortably, yet it was still hidden from view because it was so resilient, soft, and painless, that a user could forget they were even wearing them. Initial sales were

FIGURE 2.7. George H. Wilson, owner of Wilson Ear Drums Co., one of the most successful artificial eardrum companies (1922). Photograph by John T. Berry. Herald Post Collection (1994.18.4627), Archives and Special Collections, University of Louisville.

profitable enough that the newly formed Wilson Ear Drum Company took offices at the Louisville Trust Building at the intersection of Fifth and Market Streets downtown. A British branch of the company was overseen by D. H. Wilson, likely George Wilson's brother (or a pseudonym for George himself).

It is unclear how or why Wilson became involved in the artificial eardrum business, but he was certainly successful in his endeavor. As the increasing capitalist culture framed health as a purchasable commodity, it became easy for entrepreneurs to promote the benefits of the device as widely as possible, persuading consumers that good health could be achieved only through market capitalism. Imaginative inventors boasted that their new devices could harness the surgical benefits of artificial eardrums while bypassing the need for specialized medical care: their designs were intuitive, flexible, and far more effective than anything obtained from an aural practitioner. Buying and enjoying goods, then, became "a capitalistic form of self-definition," as historian Takahiro Ueyama argues. Health goods enabled consumers to achieve social emulation and idealized health.[53]

Despite their best efforts to regulate artificial eardrums to cases of perforation, by the end of the nineteenth century aurists lost control to the unrelenting market capitalism and the new therapeutic ethos that demanded equal access to all health commodities. The emergence of a new marketing style provided by full-service American advertising agencies launched national, bombastic large-scale advertisements that the medical profession could not compete with.[54] Like other patent medicine vendors, artificial eardrum entrepreneurs relied on lavish advertising to create a brand identity to confront the American citizen with the product as "as he read his mail, as he perused the paper, as he strolled the streets, as he traversed the countryside."[55] It was not just a product being advertised to deaf citizens: customers were being sold an ideal life of prosperity and happiness through hearing restoration, with the technical nature of hearing loss and the product itself fully explained in booklets, leaflets, and other promotional ephemera. In short, these entrepreneurs were offering deaf consumers the same thing aurists tended to do—hope of being hearing (again)—but they did so in colorful, accessible, and allegedly even transparent ways, by emphasizing the relative ease of their product. Artificial eardrums could be as easy to purchase and as comfortable to wear as spectacles tended to be.

In response to the unrelenting intrusion of commercialism, aural surgeons ceased to offer patients quick technological fixes and rather focused their profession on developing more intricate and specialized surgical procedures. This, they hoped, would enable them to attain jurisdictional control of deaf patients and—in the long run, perhaps—eliminate the patient-consumer's reliance on patent medicine alternatives for hearing restoration. Entrepreneurs of artificial eardrums thus exploited the burgeoning popular demand for commercialized medicine, however, placing numerous advertisements for their products on both sides of the Atlantic. Some of these entrepreneurs had questionable reputations and allegedly shady backgrounds, but they nevertheless managed to build profitable enterprises for themselves.

In an era before the regulation of medical devices, the capitalist approach toward treating deafness encouraged the growth and range of medical quackery and fraud across international borders. American John Henry Nicholson, a self-proclaimed "aural specialist," for instance, dominated the British advertising market for aural treatments during the late nineteenth century. With a capital investment of £1,000, he settled in London in 1885, opening the Drouet Institute for the Deaf three years later. Supposedly the institute was the London branch of a Parisian clinic founded by the mysterious Dr. Drouet, a general practitioner who died from tuberculosis aggravated by alcohol, but not before lending his name to Nicholson to establish the institute. Drouet apparently made remarkable discoveries that relieved his patients suffering from deafness, but what the discoveries were, he never told anyone except for Nicholson, who shared only bits of the secret with his employees, two "enterprising Frenchmen." One of the employees, a French physician and aural specialist trained at the University of Paris, Dr. H. N. Dakhyl, was employed to serve patients who visited the institute as well as by correspondence. Before he was recruited to the Drouet Institute, Dakhyl had a practice in the opulent neighborhood of Kensington in London; later, he was accused of exaggerating his qualifications and being a "quack of the rankest species."[56]

Establishing the Drouet Institute as his home base, Nicholson sold his "New and Improved Artificial Eardrums," the only "luxury" brand available on the market. They were constructed of magnetized steel rods with gold or silver covering, with two soft rubber discs affixed to each end and held in place with gold washers. Customers purchased a pair in a wooden box that contained a set of instructions for proper instruction, accompanied by a leaflet with scores of favorable testimonials.

Advertisements for the Drouet Institute and for Nicholson's eardrums boasted that the "Curse of Deafness" could be quashed.[57] There was "HOPE FOR THE DEAF," Nicholson proclaimed, encouraging customers to send in a letter with three stamps to receive the company's pamphlet, which included an essay on deafness outlining the causes of hearing loss, notices for money-back guarantees, and promotional half-price offers.[58] The institute allegedly spent upwards of £10,000 a year on advertising: ads on horse-drawn omnibuses, billboards, flyers, leaflets, and scores of periodicals invited deaf persons to send in a letter, upon which a "physician" (Dakhyl being one) would diagnose the nature of the hearing loss and prescribe treatment. Copies of testimonial letters from satisfied patients as well as endorsements from doctors, lawyers, editors, and "other men of prominence" glittered across the pamphlet and in advertisements.[59] Some testimonials even attested to Nicholson's skills as a practitioner: "They may talk of Koch and Pasteur in their cures, but for quiet, unostentatious results, give me Dr. Nicholson as an Aural Surgeon"—even though there was no evidence to support the contention that Nicholson had any medical credentials!

Nicholson promised a speedy cure to anyone who stopped by the Drouet Institute for a consultation. Sometimes there was no need to even visit the office, for it was easy and beneficial to diagnose and treat through the postal service. Hundreds of people wrote in and hundreds of replies were sent, nearly all suggesting the Drouet secret "cure-all" as a remedy: a "special" plaster to be placed over or behind the ears to release miraculous penetrating powers. Drops, gargles, and anticatarrhal snuff were also prescribed. As an added incentive for swaying customers, Nicholson narrated the tale of a "rich lady" who was so impressed with the care she received from him, that she later donated $1,000 so that poor deaf persons could obtain similar benefits. Yet Nicholson's advertisements fluctuated drastically with respect to both the story and donation amount: the donation was listed at £5,000, $10,000 and even $100,000. And as part of the rich lady's condition for donation, Nicholson's artificial eardrums were available for "free"—but only if customers forked over 13 shillings for "medicine," a lotion containing a solution of glycerine, ether, and morphia in which to wet the artificial eardrums before insertion.[60]

By 1889, Nicholson had supposedly sold the Nicholson's Patent Ltd. Company to shareholders for £100,000. Shortly after, questions emerged surrounding the integrity of his advertised claims. Henry Labouchère, the editor of *Truth* magazine, published an exposé denouncing the innovator as a "rascal" and "fraud" whose advertisements were full of malicious lies that aimed to do nothing more than defraud the deaf public, many of whom were misled by Nicholson's promises of "free" artificial eardrums. A shrewd entrepreneur, Nicholson distanced himself from accusations of fraud and quackery by establishing a new company, the Nicholson Aural Institute in Gunnersbury, and set up a fake press agency to insert his advertisement in leading English newspapers. The new company's letterhead included a likeness of Nicholson, a technical illustration of the artificial eardrums in position and a list of all the patented seals the device received from Britain, Canada, the United States, Australia, and several European countries. Nicholson also reached out to the colonial press, inserting advertisements in Melbourne newspapers boasting that his "cure" could correct deafness of forty years standing. The "HOPE FOR THE DEAF" advertisements were also circulated, assuring customers that Nicholson's Improved Patented Artificial Eardrums were accredited by the most scientific men of Europe and America, as these devices were the "only sure, easy, and unseen devices to permanently restore hearing and to cure noises in the head."[61]

Nicholson's scheme did not last long. In December 1901, he was convicted at Old Bailey in London for fraud and sentenced to eighteen months of hard labor. He eventually disappeared from public view.[62] As soon as he left, others took the spotlight. "Professor" Keith Harvey, the alias of Eddie Marr, also known as W. Scott Hamilton, who sold his fraudulent "Electrical Aural Battery," and Herbert Clifton, who supposedly cured himself with a series of lotions and his Murray Ear Drum. A. N. Wales drew connections to the Victorian marvel of electricity and sound technologies with his "sound discs," which were "the same to the ears as glasses are to the eyes."[63] The most notorious of all was American homeopath Hawley Harvey Crippen, the quiet "Chief of Medical Staff" who answered letters of people writing to the Aural Remedies Company in London (the renamed Drouet Institute) while conspiring with his lover Ethel Le Neve to murder his wife Cora. Caught by wireless telegraphy—the first suspect to be so—while escaping to the United States, Crippen was eventually hanged for his crime.[64] These men were all, in the words of Evan Yellon, the pseudonymous editor of *The Albion Magazine*, "a tribe of rat-men—the petty swindlers of afflicted people."[65]

Good Hearing a Joy

The emerging capitalist marketplace fostered a new approach to commercialized health, one that protested against restrictions on personal choice and autonomy regarding treatment. This was the rhetoric adopted by artificial eardrum entrepreneurs: they addressed their consumers' desires for robustness by insisting that they had devised a cure that even aural surgeons could not promise. Advertisements declaring "hope for the deaf" not only claimed to advocate user autonomy over control of the device, thereby stripping it of its need for medical surveillance as aural surgeons demanded, but also asserted the cultural view that deafness demanded to be cured. By purchasing artificial eardrums, deaf persons could prevent themselves from becoming "oral failures," or, in the extreme, "defective, deviant, and even un-American," as historian Susan Burch writes.[66] By promising "relief," artificial eardrums offered a chance for deaf users to appear "normal." Since good hearing was a "joy," as manufacturers claimed, it was the duty of the deaf to lessen the burden on their hearing counterparts by adopting measures to better manage their self-presentation and make conversational settings comfortable.

Eight years after he received the first patent for his artificial eardrums, Wilson's company was in financial trouble. Sales had stalled from competing variations on the market. Like Wilson's Common Sense Ear Drums, these products were advertised with the hallmarks of scientific ingenuity: illustrations featuring a cross-section of the ear with the device easily in place, boasting of superior materials, and branded with references to telephony. The newness of Wilson's design and of other artificial eardrums on the market reassured skeptical customers by offering undefined potential and guarantees of their money back.[67] It also meant that companies faced more aggressive competition in this new century.

For years, Wilson contracted the company Lord & Thomas to design his advertisements. Lackluster sales due to competition disappointment him, however, and as he debated leaving his contract with the company in 1900, he was approached by Albert Lasker. A young and ambitious advertiser with Lord & Thomas, Lasker was eager to raise his profile in the company; he and his partner, newspaperman Eugene Katz, thus proposed a new ad copy they claimed would drastically increase the Wilson company's sales. A year later, the tone and format of Wilson's advertisements transformed, shifting focus from the technical and scientific to the cultural perception of the deaf as morose and suffering persons. These advertisements reminded deaf customers that they were suffering because they were isolated in hearing society and that despite countless disappointing cures provided by medical practitioners and patent medicines, there was no device that was effective—except Wilson's Common Sense Ear Drums, made by a trusted company with over ten years in business. These eardrums were not only perfect designs of science, but a permanent cure that went beyond promises of hope. They carried the potential to eradicate all suffering associated with deafness. "DEAFNESS CURED," the new advert headlines proclaimed.

These advertisements revived Wilson's company and Lasker's

career. Most pre-1920s advertisements retained the quality of announcements to provide objective information about the product.[68] By making ad copy of news stories that appealed to consumer anxieties, Lasker created a new style of advertising art, one that "addressed people's natural skepticism regarding advertising by showing the reason why they should be interested in the product."[69] People would be naturally drawn to a story rather than a technical drawing, especially when a human connection was present that reflected their own feelings and attitudes about health. To boost the sympathetic message in the Wilson adverts, Lasker photographed a graphic artist from Lord & Thomas who looked like the "deafest man you ever saw," cupping a hand to his hear. "YOU HEAR!" the headline splashed underneath the photo, "when you use Wilson's Common Sense Ear Drums." More so, these were the only "scientific sound conductors" available; they were "invisible, comfortable, efficient," and far superior to other artificial eardrums on the market—even doctors recommended them, the adverts declared, though there is little evidence to support such claims.

The message was simple: a morose, suffering deaf man transformed with the use of Wilson's Common Sense Ear Drums. The face of the anonymous graphic artist emphasized the psychological benefits of the device. One copy included a photo of the artist, his face turned away from the camera, cupping his hand and straining to hear; the other depicted the man staring straight ahead, hand only slightly cupping his ear. If placed together, these two photos appear as a remarkable before-and-after transformation, proof of the Wilson's Common Sense Ear Drums' superiority over all other artificial eardrum variations. These new advertisements described an appealing success story to those who did not reside in Louisville or know Wilson personally—the story of Wilson as a deaf man now cured by his own creation.[70] The graphic artist, the ads suggested, was Wilson himself; if he had been cured, then so too could deaf persons across America. Indeed, one advert boasted over 350,000 deaf people cured with the device. Since Wilson was not hearing impaired, the advertisements were misleading, but Lasker's strategy worked. The Wilson company's profits increased substantially enough that Wilson tripled his advertising budget of $24,000 with Lord & Thomas.[71]

The psychological focus of Wilson's advertisements reinforced how crucial the self and identity had become; such advertisements opened the body to various kinds of commodity interventions. And as historian Thomas Richards explains, the new commodity health culture "fragmented the self by reducing selfhood to a series of acts of consumption, and it told consumers that the only way they could sustain a secure sense of selfhood was to consume more and more commodities."[72] The advertisements Wilson placed in *The Century Illustrated Monthly Magazine* certainly highlight the importance of a secure selfhood and more broadly, the cultural limitations and expectations of deaf people in America. Advertisements in this magazine were expensive, for they had to be redrawn or photographed to fit with the illustrative specifications of the editorial style. Wilson certainly could afford it. His adverts emphasized how the tendency of deaf people to withdraw into their world of silence was dangerous for them. One advertisement illustrates a woman walking on the street, unaware of the presence of an oncoming trolley, thus placing herself in direct danger. This image eerily captures an incident in which Wilson's mother, Katherine Wilson, was struck by a trolley car while she was attempting to cross the street and eventually died from her injuries; there's no indication she was hearing impaired. Claiming the car was being run at a reckless speed, Wilson filed suit against the Louisville Railway Company and in 1903 received $20,000 in damages for the death of his mother.[73]

Another one of Wilson's advertisements in *The Century Illustrated* emphasized the importance of conversation as a necessity in polite society. The photograph features three young men in suits gathered in a circle; two are laughing while the other one is hunched over, his hand cupping his ear as he strains to ear. The copy asserts that he has "Lost the Point and pleasure of conversation."[74] But with Wilson's Common Sense Ear Drums, the man would no longer be left out; he would be able to participate in ways that were previously denied to him—these devices were the key for social betterment. And more importantly, they were easily adjustable, comfortable, and invisible.

George P. Way marketed his own brand of artificial eardrums with a strategy similar to the Wilson company's. In the late 1890s, Way established his Artificial Ear Drum Company with its office headquarters inside the Majestic Building in downtown Detroit. He sold his patented, bell-shaped artificial eardrums by asking customers, "DO you want to Hear?" with advertisements featuring a photograph or illustration of his face as evidence of happiness and success with a cure of his own creation after spending twenty-five years deaf. Indeed, these advertisements embody the American democratic ideal that accentuate prestige derived from merit and expertise, rather than innate social class.[75] Way not only overcame his hearing deficiency, but supposedly also improved his speech. Here, the Way Common Sense Ear Drum claims to bypass "oral failure" by (re)connecting their deafened consumers to various sounds—"Do You want to Hear . . . the conversations of your friends—music—singing?" Moreover, the device was invisible, easy to apply, and, as mentioned in the ad copy, patented and protected, thus asserting to customers its legitimacy as an invention. They were sold for $5.00 a pair and could be ordered through the mail. For another $5.00, customers could also purchase Way's "Blowena," a cure for catarrh (an inflammatory condition causing excessive buildup and discharge of mucus in the nose/throat).

The success of the company, however, owed not to Way's business acumen, but that of his wife, Frances M. Way. Originally from Canada, the couple immigrated to the United States after marriage to secure better prospects for George's career and to raise their child, Teresa. The Way company's first patent, the bell-shaped artificial eardrum, launched the business; the patent was jointly assigned to George and William H. Seldon Jr. George received another patent in 1906 for minor improvements in design. Not two years later, Frances applied for another patent on behalf of the company, playing a sig-

FIGURE 2.8. Frances M. Way took over her husband George's business after his passing. She was one of the few women entrepreneurs of artificial eardrums (c. 1900s). Erickson Family Archive, courtesy of Carl Erickson.

nificant role in its success by transforming the product into a medical device to better defuse accusations of fraud.

A 1902 graduate of the Detroit Homeopathic College, Frances was among the first women to practice medicine in Michigan. She focused her studies on understanding defects of the ear, which she then applied towards creating a superior version of the original Way eardrum. She added medical preparations to help support insertion, thereby transforming the ordinary artificial eardrum from an acoustic aid into a medical product, advertised as a "medicated ear-drum." The company's advertisements identified George Way as the inventor of the Way eardrum; Frances is referenced only as a "specialist" who had verified the medical benefits of the product. In 1919, George died suddenly from a stroke while attending services at the Central Methodist Church. Frances took over the reins of the company and assumed most of George's responsibilities. Detroit businessmen recognized her entrepreneurial spirit and respected her role within the company. Four years later, she filed for and received trademark registration for the Way Common Sense Ear Drum, although the *Official Gazette of the United States Patent Office* listed her as "doing business as George P. Way."[76]

Frances Way was not the only woman entrepreneur who sold artificial eardrums. Laura H. Vickers of West Philadelphia received a patent in 1900 for an artificial eardrum branded as the Morley Ear Phone. She married George M. Vickers Jr. in 1895 and a year later, the couple had their first child, George M. Vickers III. Daughters Beatrice and Elizabeth soon followed, joining the rest of the family at their dwelling in Philadelphia. Unlike Frances, however, when Laura received her first patent, she assigned it to her husband. The Vickerses also founded the Morley Pharmaceutical Company and operated it out of an office in the Perry Building on Chestnut Street, selling the artificial eardrum primarily through mail-order advertisements.

The Vickerses' product was branded with the tagline "Don't Shout!" since their "telephones for the ear" allowed even whispers to be heard, thus elucidating both the century's fondness for communication technologies and the common perception that raising one's voice was an appropriate solution to conversations with a deaf person. A 1903 advertisement, for instance, illustrates a deaf person seated in a chair, raising a hand to another individual and saying, "I hear you. I can hear you as well as anybody. No need to shout." The Morley Ear Phone guaranteed a perfect fit for up to two years when worn day and night with comfort, safety, and satisfaction. "It returns hearing like magic. A child can use it. When adjusted it is invisible. No one can see it."[77] Anyone with poor hearing can use it, except for people born deaf—for that would be equivalent to putting glasses on a totally blind person.

Deafness Fakery

"I am hard of hearing and want help," wrote F. S. Fount to Arthur J. Cramp, the first director of the American Medical Association's Propaganda Department (later renamed the Bureau of Investigation). "But have found nothing in the way of an ear drum that was any good." Were there any other artificial eardrum models that the AMA could recommend, Fount asked, especially ones that could cure his deafness in a timely and inexpensive manner?[78] Patent medicines were a big business and artificial eardrums were no exception. The sheer number of advertising campaigns launched by prominent companies indicates the extent to which deaf people were exposed to artificial eardrums as an easy, permanent correction for hearing impairment. Moreover, deaf periodicals and medical journals alike discussed the merits of these devices, further promoting the potential possibilities of a cure and reinforcing its paradoxical status as a device "that existed both outside of and within mainstream medicine."[79] Yet among the strategies and techniques of persuasion used by advertisers to sell these devices lay the likely fact that artificial eardrums were mere gimmicks, useless to all cases of deafness save eardrum perforation, and that their benefits had been grossly exaggerated by charlatans hoping to make a quick profit.

Established in 1913, Cramp's department aimed to gather and disseminate information to the public and profession regarding health fraud and claims of quackery, beginning with the resolution that the AMA would refuse advertising space in its journals for unverified or unapproved medical preparations. The organization later expanded its campaign against quackery by declaring that all patent medicine advertisements were objectionable. Yet as historian Eric Boyle reveals, even as the AMA managed to eliminate patent medicine advertisements from medical journals and some lay periodicals, other print materials still made their way to doctors and consumers through the mail.[80] Thus, the Bureau of Investigation requested that all products accused of being fraudulent be reported for investigation. If a thorough examination of the product's advertising claims and health benefits revealed it to be misleading or fraudulent, then the report was published in the AMA's journal and forwarded to the appropriate channels to bring forth criminal charges. The most notorious products, the get-rich-quick-schemes, the huckster hokum, the fakes and fraudsters, were later compiled by Cramp in his five-hundred-page volume, *Nostrums and Quackery*.

Deafness cures especially came under fire. Cramp listed all artificial eardrums as useless and ineffective, citing British writer Evan Yellon's exposés that concluded that they "are dangerous things to use, and it is madness for any unskilled persons to meddle with the ears."[81] Echoing the insistence of otologists, Cramp noted that most artificial eardrums were unnecessary as aids to hearing and that only a small percentage of deaf people actually benefited from the devices, which were also easily obtainable upon diagnosis and prescription from qualified medical practitioners. For all other cases, "indiscriminate sales of a device of this sort, especially at exorbitant prices and under false and misleading claims is not merely an injury to the purse, but a distinct menace to the health of the deaf." Touring the country to deliver presentations on deafness cure fakes, Cramp told audiences of otologists, social workers, caregivers, interested family members, and hearing-impaired people that while artificial eardrums appeared to be a simple, even exciting solution, too many people fell victim to quack sellers, swayed by fanciful promises of a permanent cure. In one instance, he referred to Yellon's interview of a deaf man who visited *The Albion* office with a small tin box containing every brand of artificial eardrum he purchased, including Cousin's "hat-shaped" membrane, Wilson's Common Sense Ear Drums, and a pair of anonymous "Audiphones."

Exposing suspected quack frauds was a difficult endeavor. Investigations often spanned several years and at times involved collaboration with other organizations, such as the Post Office or the Federal Trade Commission. As early as 1910, the Bureau of Investigation inspected Asbury Osman Leonard's "Invisible Antiseptic Ear Drums," which were initially peddled in 1907 as an accessory to "Leonard Ear Oil," then later marketed as a stand-alone cure. A form letter from Leonard boasts of his focused approach to sales, including that his products are handled by him only. "Unseen Comfort," he claimed, was a concept he intimately understood, for he too, suffered the "same embarrassments and loneliness" that his customers did.[82] "You want your Hearing Restored and Head Noises stopped," Leonard urged his customers. "Therefore you must do something to secure relief. I claim that my Drums and Oil are accomplishing these results more successfully than any other known means and I challenge any one to dispute this statement." To obtain maximum benefits, customers were instructed to wear the rubber eardrums and dab the oil behind their ears, to ensure that the medicinal preparations would be effectively absorbed by the skin to cure deafness.

Following up on some of the testimonial letters advertised by Leonard's company, the bureau concluded that some were fake, thus denouncing the product and ordering all advertisements to cease being printed in periodicals. Moreover, Leonard allegedly sold the original letters he received from prospective customers after they ceased to be commercially valuable for his business. He eventually abandoned the mail-order business and his eardrums, selling only the oil through retail drug channels and splitting profits with druggists, much to the dismay of medical practitioners. One physician, Dr. Robert Harry Scott, wrote a letter to Cramp mentioning that he "happened to be standing in a drug store and the druggist tells me that [the Leonard ear oil] is a great seller. $1.00 for a small bottle and rub it in back of the ears I think would clean out a purse long before it would clean out the Eustachian tubes."[83] On 14 July 1918 Leonard was arrested in New York City on complaint of the Department of Health on grounds his claims for the oil were false and misleading; the "medicinal ingredients" included in the oil were nothing more than an emulsion of mineral oil, soft soap, glycerine, and eucalyptol.

Arraigned before a Court of Special Sessions, Leonard was judged guilty and sentenced to thirty days in jail or a fine of $250—he paid the fine. The department also notified all druggists to stop the sale and promotion of Leonard Ear Oil, explaining that any disobedience of the order would make the druggists liable to prosecution. The Cleveland Health Department issued a similar order, though Leonard kept operating until 1936.[84]

People still hoped that an advertised artificial eardrum could cure their deafness. Hundreds of concerned deaf citizens across America wrote to Cramp, asking, even begging, for advice or validation in their search for a cure. Family and friends inquired whether their loved ones were wasting their wages on an advertised eardrum, and physicians asked—sometimes on behalf of patients—whether the latest deafness cure had any medical merit. These letters all pleaded for hope. Others expressed healthy skepticism. "I am reluctant to send an order for the drums unless I have information from a good reliable source that they are worth the investment," James S. Mills wrote in 1915.[85] Why wouldn't he be? The cost of a pair of artificial eardrums, as sold by qualified aurists, were no more than one shilling in Britain, or seventy-five cents in the United States. The average costs of materials to assemble the device was about twenty-five cents. The enormous markup of advertised eardrums—costs ranging from $5.00 on average to the exorbitantly priced $10.00 Actina, a small steel vial with a screw stopper at one end, advertised for catarrh deafness—was not only ridiculous, but victimized an already vulnerable public. "I am victim of the deafness will [*sic*] ask your advice before investing" in Way's eardrums, one citizen wrote.[86] "I am going slowly deaf & don't know & even specialists don't seem to know the cause. . . . I have purchased three machines since I got this way & am seriously thinking of purchasing another," confessed A. P. Hilton.[87] "Is there not some way that such people can be prevented from duping the aged and ignorant?" wrote concerned citizen Alvin P. Cameron. After all, "Ears are too delicate and important human assets to be placed in the power of quacks."[88]

George W. Prohaska of Iowa wrote a letter to the bureau dated 27

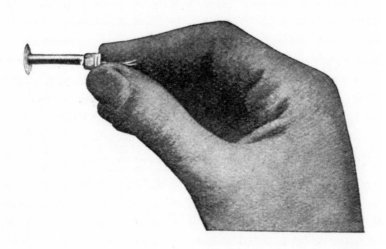

FIGURE 2.9. Close-up of the Morley Ear Phone, a type of artificial eardrum (1910). Pamphlet, Hagley Library.

October 1927, asking for expert advice as to whether the "use of any ear phone say the Moorley's [*sic*] or Leonard Ear Phone replace the true ear-drum as to hearing qualities?"[89] After examination by specialists in Rochester and Des Moines, Prohaska received a diagnosis that his hearing loss was due to a severe case of whooping cough and there was nothing to be done. Unsatisfied with the result, he visited another specialist, an osteopath who diagnosed his hearing loss as being due to bad teeth and tonsils; the osteopath told Prohaska his eardrums were getting "hard like stone," white in color, and puncturing the eardrum was the only way to improve his hearing, or at the very least, stop the "ringing" noises in his head. Thus, Prohaska implored, will artificial eardrums help?

Similar letters inquired whether different models of artificial eardrums were advantageous, sharing their stories of hearing loss and struggles with finding a cure or treatment. In his replies, Cramp advised citizens to forgo artificial eardrums, directing them to *Nostrums and Quackery* or the bureau's pamphlet, *Deafness Cure-Fakes*, which contained investigative results and a listing of companies found to be fraudulent. The Morley Ear Phone was nothing more than a modification of Toynbee's rubber disc design, though it had "the effrontery to say that this is the 'latest and most effective device to the relief of deafness.'"[90] The adverts and seventeen thousand letters eulogizing the Ear Phone were misleading, as were those for Way's eardrums. Discrepancies in the company's advertised statements were exposed, and George Way's claim of receiving foreign patents was debunked, as was his assertion that he was deaf for twenty-five years before being "cured." Even the advertisements for Wilson's Common Sense Ear Drums were criticized for making superlative claims, as the product was declared "just as worthless as most devices of this sort and just as potent for harm"—though Wilson dispatched his lawyers to help clear his name against what he perceived as a biased and unwarranted attack on his company. Even E. E. Bullis of Lord & Thomas was shocked to discover his client was accused of fraud, as when they handled the account "it was a prosperous company doing big, legitimate business."[91]

An array of letters, all littered with pleas for hope. "In your article," wrote Judge Thornton W. Sargent of Wichita, Kansas, to Cramp, "you state that [Way's] ear drum cannot CURE anyone. Of course, it cannot cure anyone but I would like to know if it can HELP one's hearing."[92] Finding Audiphones and Acousticons (early carbon electric hearing aids) too inconvenient because of the wires and pieces, Sargent hoped Cramp could advise on a proper eardrum that would help him continue his work at court. Leonard F. Weiss sent a letter asking Cramp for advice on what make of earphones could help him regain his hearing, which was lost at age eight. Now twenty-four years old and stone deaf, he was debating between Mears Ear Drum, Leonard Ear Oil, Megaphones, or Wilson's Ear Drums.[93] "My sister is using [Way's] and seems to have faith in them," writes Elise Wicken in a 1930 letter. "Tho we notice no improvement and she has had them about two months. Could the medicine that she puts on them before inserting them in the ear be harmful and make her hearing worse[?]"[94] Again on Way's: "I am getting deaf & and understand that any a number of Invisible Ear Drums on the market some of which on faith & some that have some merit."[95] For seventy-three-year-old T. A. Casey of Alabama, a Sonotone bone conduction hearing aid was unsatisfactory for a "slight impairment of hearing," and thus, wondered whether Way's eardrums were a better choice.[96]

As the *Deafness Cure-Fakes* pamphlet received greater circulation, leaders of the deaf community condemned advertised cures for deafness, urging deaf persons to refrain from purchasing these fraudulent and likely dangerous goods. The *American Annals of the Deaf*, a periodical catering to d/Deaf readers, produced a list of artificial eardrums to be avoided, including the Morley Ear Phone, the Way eardrum, the Wilson eardrum, and other patented "cures" for deafness: Actina, Branaman Remedy, Coutant's Cure, Gardner's Cure, and the merchandise of Dr. Grains Company and the Help-to-Hear Company.[97] Likewise, John Dutton Wright, founder and director of the Wright Oral School in New York City, warned parents of deaf children that "there are many persons who live by trading upon the ignorance and credulity of the unfortunate. The deaf and the friends of the deaf fall an easy prey to the advertisements of quack remedies, ear drums, etc., that are always useless and sometimes actually dangerous."[98] He advised deaf persons, and especially parents of deaf children, to ensure that they were informed of fraudulent products on the market and not to be swayed by their colorful claims of a miraculous cure.

Advertised artificial eardrums fell out of favor after the 1930s, in no small part due to the campaigns led by the bureau, the Post Office, and the FTC to eliminate their quackery. Better technological excursions came along as well: newer, more powerful hearing aids that could be used for a wide array of auditory limitations and safer, though questionable, surgical maneuvers.[99] Not all people, however, accepted the dismissal of artificial eardrums. Much like Harry Stephen Keeler's deafened burglar, some people insisted the devices were more helpful for aiding their hearing than any other available product. In 1948, for instance, Gisella Selden-Goth of New York City returned her copy of a popular book, *Hearing and Deafness: A Guide to Laymen,* to the publishers, because it failed to include any discussion of the benefits of artificial eardrums. She argued that otologists who dismissed the device's value were fostering prejudice against "one tenth of a percent of America's hundred-thousands hard of hearing suffers [who] have been helped by artificial eardrums."[100]

Even George Wilson appeared to have stayed in business. Though no advertising materials for the Common Sense Ear Drums appeared in the 1930s, in 1938 he claimed to have created "a new and improved Ear Drum far superior to any [he had] made in the past." He also remained bitter against the American Medical Association for hounding his enterprise, arguing that every company that dared to interfere with the "Doctors business" during the Depression were especially attacked for "getting all the grapes."[101]

Despite newer technological developments in electric hearing aids, the stigma of deafness persisted. Even as deaf users embraced their prosthetics as ingrained extensions of their body and of their identity, the twentieth-century commodity culture introduced more technological fixes designed to disguise deafness. While eyeglasses transformed into fashion accessories, hearing devices remained stigmatized. Indeed, even the design of hearing aids into eyeglasses frames—a new twist on "ear spectacles" that first emerged in the 1950s—did little to change the cultural expectation that deafened persons correct their infirmity and become hearing, or at least pass as such. Yet these devices propagated the same messages as artificial eardrums did: in 2018, U.K.-based Hidden Hearing, for instance, highlighted the fact that their "Spectacle Hearing Aids" are "discreet—no one need know you are wearing a hearing aid."[102]

3

Electric Wonders

It was pathetic to hear grown-up men and women repeating with delight "Papa," and "Mama," and "hello," the meaning of which was explained to them by movement of the lips, which all seemed to understand.

Boston Globe (29 August 1900)[1]

I have long felt, that if someone who understood, cared enough to really try, something might be invented, that would do for the partially deaf what glasses do for the partially blind.

Lucy Taylor, *The Silent Worker* (1913)[2]

When she sang, her voice was agile and pure, drawing rapturous applause across world stages. For Orris Benson, Suzanne Adams was a marvel. Not only for her lovely timbre, but for her willingness to participate in this demonstration with him, so that he, who had never heard sounds—let alone a grand opera soprano—could finally be able to do so. He placed headphones over his ears. Wires tangled down from his chest towards the desk, connected to a contraption, which in turn was connected to the speaker-trumpet Mme. Adams was using. As sparks traveled through the wires, Benson realized that the magic of her music was made possible for him only through the wonder of electricity. Through electricity, he could move past the barricades of his natural senses. He could finally hear music— Mme. Adams's music to start with—the language of hearts, the divine chord of humanity.

This was "curing the deaf by electricity," wrote Bennett Chapple

FIGURE 3.1. Orris Benson tests the Akoulallion with opera singer Mme. Suzanne Adams. *National Magazine* 18 (April 1903).

for *The National Magazine* in April 1903 as he witnessed the scene between Benson and Mme. Adams.[3] Electricity was becoming a familiar feature of the American cultural landscape, its potential limitless. It was once a spectacle and source of public entertainment relegated to traveling showmen and their displays of sparks and shocks; by the second half of the nineteenth century, electricity had slowly crept into the fabric of daily life.[4] It was the "wonder of the age, the hallmark of progress," a mysterious power bordering on the supernatural that reverberated through wires to electrify America and transform its landscape.[5] From lighted houses to wireless telegraphy, a range of machines were developed to harness this curious power. Telephones, trolleys, and toasters became regular features of the new electric life. The contraption Benson was hooked up to was hardly novel — many experiments in sound technologies and carbon microphones had led to this moment, especially Alexander Graham Bell's telephone transmitter — but it was certainly the most exciting.

It promised to enable at least 95 percent of deaf-mutes to hear sounds. With time and training, at least 80 percent of users would be able to hear and speak. A magnificent cure in a mechanical box.

This instrument was the Akoulallion—Greek for "to hear" and "to speak"—a tabletop carbon-fiber apparatus with a microphone. The brainchild of American engineer Miller Reese Hutchison, the instrument was initially designed for deaf-mute children to learn in a group setting. Patenting the device in 1898 under the original name "Akouphone" and manufacturing it out of his Akouphone Company of Alabama, Hutchison promoted the Akoulallion as an "immediate benefit" for deaf people. The Akoulallion also enabled deaf-mutes to hear their own voices, some for the very first time. At the 1900 annual convention of the New England Gallaudet Association of the Deaf in Boston (the predecessor for the National Association for the Deaf), Hutchison exhibited his instrument and invited delegates to try it. As a *Boston Journal* reporter noted, watching the deaf try the instrument was astounding, for as sounds crept to their ears, many did not even recognize what they were hearing. "A girl who could speak so that she could be understood," for instance, "and could read the lips of an ordinary speaker, was utterly at a loss when the actual sounds reached her. She could not correlate sound and meaning; though she could correlate lip-motion and meaning. Many—indeed, most of them—jumped at the first sound that reached them as though they had received a shock of electricity."[6]

Another reporter observed a "pretty and stylish young woman" who had never heard a sound or spoken a word since being afflicted with diphtheria at age three, anxiously watching others try out the instrument, eager for her turn. Upon wearing the earphones, "She quickly learned to coo 'papa' and 'mama' and 'hello,'" the reporter recounted, "and when the brass band was turned on and the inspiring rhythm of a quickstep was revealed to her, her face lit up like that of an angel, while her dainty gloved hands instinctively beat time in unison with the glorious sounds which until that instance she had never known."[7]

Sold in limited units for the egregious amount of $400 (approximately $12,000 in 2018 dollars!), the Akoulallion was not a market success. In 1900, Hutchison redesigned the instrument, making it smaller, practical, and portable. Priced at $60, the Akouphone consisted of a double carbon microphone set weighing 3.5 pounds and attached to an earphone; a six-volt storage battery was carried in a separate rubber case. A tremendous improvement from Hutchison's original design, the Akouphone became the first commercially available portable hearing aid for deaf citizens in the United States. It was not the *first* electric hearing aid, however; the honor for the first such invention goes to Dr. Ferdinand Alt of Vienna, who, in 1892, developed a protype earphone connected to a carbon microphone and fastened to a battery box.

Though a novelty and a testament to the capacity of electricity to deliver sound to the deaf, the bulk and the hefty cost of Hutchison's devices made them unpopular and unprofitable. For instance, writing for *The Silent Worker*, a popular newsletter for deaf readers, Alexander L. Pach accurately discussed the merits of Hutchinson's machines. He acknowledged that many deaf people felt "fooled" by the illusion of hearing "sounds" when in fact, they only felt "vibrations," which needed no machine to replicate.[8] The potential for assisted hearing amplification, however, quelled many naysayers, especially after Denmark's Queen Alexandra invited Hutchison to demonstrate his machine for her. Marveling at the Akouphone's ability to partially restore her hearing, the queen requested that Hutchison officially present his machine during the 1901 coronation of her husband, King Edward VII of England. Hutchison received a royal medal and a generous stipend as a reward for his service. The Akouphone also gained worldwide esteem. Eight years later, however, the queen's hearing worsened. Finding mechanical acoustic appliances to be of no practical use, she disregarded all of them, preferring instead to lipread, which enabled her to converse freely with her close friends.[9]

After several company reorganizations, Hutchison and his business partner Willard S. Mears formed the Hutchison Acoustic Company. In 1902, they unveiled the "Acousticon": a body-worn carbon hearing aid with a single microphone and powered by batteries lasting from a few hours to a week. Practical and resilient, the Acousticon did not provide sufficient acoustic gain for those with anything more than moderate hearing loss, but at least the quality of the instrument and its internal noise were bearable for users.[10] This third-generation model offered customers the possibility of (re)gaining "normal" hearing, especially since it was tested and even—at least partially—designed by deaf and partially deaf consultants. R. E. Maynard, for instance, notified readers of *The Silent Worker* that the improved design was better than the Akouphone, but it was still problematic in distinguishing sounds that were "heard" rather than "felt." "It felt so faint," he explained, "that no distinction could be made without the inventor first teaching the difference in sounds in the words 'papa,' 'mamma,' [and] 'hello.'"[11] While the Acousticon required further improvements to benefit congenitally deaf users, Maynard added that "it was clearly demonstrated that the device will prove highly helpful to the *hard of hearing*, and for that purpose is probably a great success." It certainly was useful for people like Lucy Taylor, who was delighted to try out the Acousticon at one of Hutchison's demonstrations. In 1913, she wrote to *The Silent Worker* that the Acousticon gave her the "first ray of hope" she had in many years.[12]

Electricity is an old business. It dates to antiquity, when electric eels were harnessed to provide electric shocks for treating people with gout or paralysis.[13] Medical electricity, however, did not become widespread until the eighteenth century. The invention of the Leyden jar in 1745, a device for storing electric charges, was one of many experiments by those who investigated the healing properties of electricity. These "medical electricians" claimed that electricity was a cheap and accessible source of therapy that required trained and skilled minds. They created new technologies for the therapeutic administration of electricity and presented accounts of their medical cases to their profession. As historian Paola Bertucci argues, these cases were necessary for "providing trustworthy testimonies and calling for virtual witnessing," especially as these healers attracted both skepticism and indifference from the medical establishment.[14] By the 1770s, electricity had become a proper medical therapy administered by medical electricians and physicians, but the general fascination with shocks also made it perfectly fitting with the self-help ethos of the time.[15] Electricity then, became a fashionable way to stimulate the body to health. Even Beethoven tried shocks delivered through water and machines to cure his deafness.

Although the early electric machines were initially marketed only to physicians—being unsuitable for home use in untrained hands and too expensive to sell profitably—the range of appliances that emerged in the nineteenth century by so-called "electric entrepreneurs" centered on the ability to cure beyond the physician's office and into the home. In some instances, these machines were purchased by wealthy citizens and installed in their homes for direct access to electricity at any time and on any day they required; other machines were carried by the trained (or in some cases untrained) practitioner to the homes of their patients. This idea of electricity as medicine, referred to as "electrotherapy," capitalized on the idea that if messages could be sent through wires over long distances, then perhaps electricity surging through the body could "unblock" diseases to improve health. Thus, as electrotherapeutics began its ascent during the nineteenth century, a staggering array of machines appeared on the market, promising immediate benefits for a host of diseases and disorders. Electrotherapists claimed electricity was a needed relief against ineffective nostrums or heroic drugs that wreaked havoc on the body. Promotional paraphernalia for these machines certainly made grand proclamations: "Use Common Sense Instead of Drugs." "Health Without Drugs." "Electricity Is Life."

"Electric health was a big, and lucrative, business," historian Iwan Rhys Morus tells us.[16] Cures were placed at what Carolyn Thomas de la Peña calls the "center of a dizzying, modernizing electric age," the machine shocks releasing "blocked" energy, reinvigorating muscles and revitalizing nerves in an inexpensive, easy-to-use manner.[17] Michael Faraday's discovery of the induction current in 1831, which provided the first continuous electrical current, quickly ushered in the production of machines that converted mechanical to electric energy. With continuous current, the electric entrepreneurs promoted electric energy as a means of salvation, an extraordinary force that called on the power of nature to establish health. They banked on the consumer's quest for a healthier and more vigorous life, creating apparatuses that could transfer "bad" electricity out of the body and return an appropriate quantum of "healthy" shades, simple enough for the consumer to administer self-treatment without complications. These proprietors sold galvanic belts and Pulvermacher's chains, strips of copper and zinc, Renu-Life violet rays, ozone generators, electric baths, and handheld massagers, all designed to restore virility and vitality. These machines, with their complicated switches and levers, became black boxes full of promises that shifted from fantasy to reality as the mighty healing power of electricity coursed through the body.

Then the electric age made electricity a spectacle. Thousands of electrotherapeutic machines were manufactured, flooding the marketplace as savvy capitalists extolled the virtues of their wearable electric devices—some which were medicinal, others pure novelty. Some were nothing but scientific gobbledygook. Most machines relied on batteries or produced electricity through a simple chemical or magneto-electric action; others used an induction coil, sometimes with a battery, for a smoother and higher intensity charge.[18] There are three forces of nature embodied in these machines, the electrical entrepreneurs claimed: vibration, massage, and electricity. Together, their natural power could defeat any incurable disease, harnessing the guiding force that governed the universe to metaphorically serve as the right arm of God.[19] Never mind that the black boxes were sometimes empty or stuffed with unmovable dials and levers that were more decorative than functional.

As historian Nancy Tomes has argued, nineteenth-century advertising and marketing efforts aggressively promoted all kinds of products, including electrotherapeutics as "essential aids in the American consumer's quest for good health."[20] Advertisers rivalled physicians as "purveyors of popular knowledge," encouraging the public to self-diagnose their conditions and purchase one of the many curious electrical fads that began to flood the market.[21] Electric belts were especially popular: a lack of nervous energy, proprietors claimed, weakened the body, and thus, recharging it with a jolt or steady stream of energy could revitalize it. Many of these devices were junk, marketed by charlatans aiming to make a fortune from their pseudoscientific creations.[22] Yet they fortified the perception that commercial electrotherapy devices were *better*, perhaps more trustworthy, in a consumer's hands, than a visit to the doctor. And even if consumers did not understand the principles of electrical medicine, they at least understood the cause-and-effect logic of sickness and cure: they could see and feel electricity at work. Those who purchased an electric device were not acting out of ignorance, but seizing an opportunity. They were not embarking on an unreasonable and irrational path to health, but strolling along a resourceful trail of health care.

The Acousticon, like all portable electric hearing aids that followed it, joined the list of electrical machines, gadgets, and gizmos that were emblematic of the American march of progress that perceived technological advancement as part of a glittering new modern reality. As *Cosmopolitan Magazine* reported in 1912, electricity brought "a new capacity for transmitting sound and a new wonder for magnified sounds," enabling engineers to incorporate the "true scientific principle of the telephone" to devise more powerful acoustic aids. References to hearing aids as "telephone aids" or "telemicrophone aids" extended a familiar communication technology into the realm of assistive devices. There were branded versions of this, such as the Gem Phone. References to the telephone also invoked a need to normalize and destigmatize hearing aids, making them a personal form of the household appliance. Or, as in the case of Montana's J. C. Chester, telephony could represent the ability to transcend barriers between man and machine. A deaf-mute who learned to speak but could not rid his speech of the guttural sounds peculiar to deaf speakers, in 1897 Chester fit a telephone to his body to amplify sounds and make conversational bearable over a space of several feet. In his quest to improve situations generated by his deafness, Chester literally made himself into a "living telephone" and marched himself to Washington D.C., to patent himself.[23]

Hearing aids, however, could only amplify sounds, not necessarily restore hearing. They were—and are—*not* cures. They were an improvement over the uncomfortable mechanical aids of the nineteenth century, especially after engineer Earl Hanson patented the first vacuum-tube hearing aid in 1920 and paved the way for smaller and more powerful designs, leaving the time of ear trumpets, horns, and eardrums in the past.[24] But they still failed to give genuine satisfaction for consumers. Noise interference, conspicuousness, and short battery life frustrated most users, especially users who purchased a hearing aid believing that they were buying a permanent cure.

Perhaps skeptical of their products' profitability as an assistive device and wanting to address customer demands, hearing aid companies began to sell additional electrical devices as a "cure" for deafness. This became an attractive marketing strategy to retain sales from previous customers who were disappointed with the performance of their hearing aids, or those with severe hearing loss. These proprietors—who would later in the twentieth century become some of the giant hearing aid companies—sold their own brand of electrotherapy gadgets, vibrators, and massagers, often at discounted costs with the purchase of an aid. This was, after all, the age of electrotherapy, when mechanical devices were not restricted by the 1906 Pure Food and Drug Act. Indeed, electrical entrepreneurs could boast fantastic claims of efficacy for their machines, and they did so until 1938, when the sale and promotion of mechanical devices came under federal regulation.

These electrotherapy devices for deafness, the vibrators and the massagers, did not necessarily follow a linear progression that diminished with the emergence of electric hearing aids. Perhaps more importantly, the fact that hearing aid firms were selling questionable devices is indicative of the cultural stigmatization against deafness, such that a "cure" remained highly desirable even as assistive technologies were becoming useful for hearing-impaired people to manage their everyday lives.

Caroline and Carolyn

Electricity, medical electricians asserted, was the key for restoring the body's finite quantity of "nerve force," which tended to be depleted in a fast-paced and bustling modern society. Physicians testified that modern life was responsible for an array of chronic symptoms observed in patients: depression, irritability, insomnia, lethargy, and pain.[25] These were symptoms of a disease defined in 1869 as "neurasthenia"; an inability to adjust to the turbulent responsibilities the modern citizen was expected to bear. Modernity caused nerves to deteriorate, leading sufferers to find life depressing and unpleasant. The diagnosis gave Americans good reasons for using machines to rejuvenate their bodies, these anxious bodies that required external power sources to keep functioning. In some instances, electricity was perceived as the only way to restore the nerve force, or even for harnessing "damaged" or "dead" nerves in the ear, as was the case for Carolyn Perkins of Rochester, New York.

Carolyn Erickson Perkins was born on 24 July 1868 to businessman Gilman and Caroline Perkins; Caroline was the daughter of Aaron Erickson, founder of the Union Trust Company. At the age of nine or ten months, Carolyn had an attack of laryngitis. For three days, she lost her voice and was unable to cry or make a sound. She recovered swiftly, and her parents thought nothing more about her illness until she began to have severe crying spells. Thinking the cause to be infant colic, Caroline fed her daughter anise tea to help ease the spells. By the time Carolyn was two years old, her parents suspected something was peculiar about their daughter's development. She was restless, only imitated certain words, and had not yet learned to talk properly. Nor did she turn when called, unless she was sitting on the floor.

After the family physician suggested that Carolyn was deaf, a specialist by the name of Dr. Rider was summoned to confirm the opinion. Much to the Perkinses' dismay, Rider agreed with the physician. Dissatisfied with the diagnosis, the Perkinses took their daughter to be properly examined by renowned surgeon Cornelius Rea Agnew at the New York Eye & Ear Infirmary. "He knew," wrote Caroline in a 28 December 1897 letter to another specialist, H. A. Tannous, "from her squirrel like movements, before he had made any examination, that she could not hear. He said it was hopeless, and we need never to try to do anything for her, that she . . . had no auditory nerve, or it was paralyzed."[26] Distraught, the Perkinses accepted Agnew's assessment. In later years, however, Caroline blamed herself for failing to disclose Carolyn's full medical history to Agnew, namely, the laryngitis; instead, she explained to the surgeon that Carolyn's deafness was likely the result of "shock" from a severe thunderstorm.

When Carolyn was six or seven years old, the family traveled to Maryland to visit relatives. There, Caroline decided to tour the Maryland School for the Deaf, where she met teacher Mary Hart Nodine. Impressed with Nodine's work, Caroline coaxed her to come to Rochester and work as Carolyn's private instructor. As the Perkinses were aware, no formal schooling was available for deaf children in Rochester, the nearest opportunities being either the New York Institution for Deaf-Mutes or the Hartford School in Connecticut. A year later, in February 1876, the Perkinses convinced local Rochester businessmen and Mayor George Clarkson to establish the Western New York Institution for Deaf-Mutes. Nodine's fiancé Zenas Freeman Westervelt became the first superintendent of the institution, which would expand to become the Rochester School for the Deaf in 1920 and remains one of the largest schools for the deaf in North America.

Rather than keeping Carolyn from public view, as many affluent families did with deaf children, the Perkinses advocated for the right of deaf children to obtain equal education, working to transform prejudicial perceptions of deafness. Caroline's writing reveals that even as her daughter learned to communicate orally, as a mother she still wished for the deafness to be cured—if not by the physicians, then by other healers. The first alternative healer was a lady

"full of magnetism" whose hands possessed powers capable of healing any ailment. Carolyn was treated daily over a period of three to
four months, until one day, the young girl complained that the sound
of low notes on the piano hurt her ears.

"She had never before mentioned a sound," recalled Caroline,
"and I felt sure that something was coming from it, but no further
progress was made, and I gave up the experiment." The experience,
however, solidified Caroline's determination: "I made up my mind
from this time that she would hear, and have never abandoned the
idea, patiently or impatiently waiting for new discoveries."

For some time, Caroline took her daughter to visit new physicians.
When Carolyn was sixteen, physician Edmond B. Angell presented
electricity as a treatment option, theorizing that the auditory nerves
were weakened and needed external charges to revitalize them.
Having gotten the Perkinses' permission, Angell devised a course
of electrotherapy for Carolyn comprising a series of mild electric
shocks applied directly into her ears. In a diary entry, Caroline notes
that the treatment commenced on 8 August 1884 and that after a
week, Carolyn heard the noon whistle for the first time, and "came
rushing up in my room in the greatest excitement to ask what it was.
Naturally she had never known about it, no one thinking to mention what she could not hear." Caroline kept notes on her daughter's
progress as well as that of Carrie Hudson, Carolyn's deaf-mute friend
who was also under electrotherapy treatment from Angell. Often,
Angell tested both girls to compare their improvement, even discovering that one of Carolyn's ears was better than the other, though
the girl hardly noticed a difference.

The electrotherapy treatment continued every day over several
weeks. Carolyn noticed sounds she never heard before: a dog's
barking, the scrape of dinner chairs on the floor, children scream-
ing at play, the moment the band at the skating rink stopped playing, and the cheers of a crowd gathered to hear a governor's speech.
After several months, Angell modified the treatment, inserting a
"little rubber tube in the form of an ear trumpet in her ear, and then

[pressed] in front of the opening a bulb which forced the air across the tube with a noise." Supposedly the procedure further improved Carolyn's hearing, as her electrotherapy treatments were reduced to three times a week.

Caroline researched other treatments for her daughter as well. She followed Boston aural surgeon Charles J. Blake's advice that Carolyn use an ear trumpet to assist her hearing as well as speech. Blake, who, alongside Alexander Graham Bell, invented the phonoautograph, an early sound recording machine, believed that the mechanization of the ear's tympanum could be aided by a trumpet; sounds vibrating from the trumpet would then reach the cochlea and enable the user to hear. Carolyn was also examined by Bell, who asserted that despite the use of the ear trumpet, the girl was not actually hearing vowel sounds, but rather repeating from rhythm. "It makes me faint over, when I think of what a blow that opinion was to me," Caroline wrote, "but I have never given up hope, and the fact that she knew so well the vowel sounds when taking electricity, it seems to be a proof that Dr. Bell was mistaken." Other treatments soon followed. Though Caroline refused physician Samuel Sexton's suggestion for a "new and experimental surgery," she did take Carolyn to Chicago in 1893 to visit "a man who had performed some wonderful cures through massage. He worked upon her head and ears for three months, without much result, and this was the last thing done."

While supportive of the work of educators for deaf-mutes and their work in teaching Carolyn speech, Caroline never abandoned the search for a deafness cure. By 1888, Carolyn's speech was improving, but she did not "make any signs," likely due to Westervelt's decision to discourage pupils to sign, believing that it would impair their ability to properly understand English. Instead, he taught them manual alphabet and speech, what became known as the "Rochester Method."[27] For Carolyn, electricity had worked for a while, especially combined with speech lessons, to enable her to "hear." "Every new suggestion for her," the mother expressed in her letter to Dr. Tannous, "makes me too hopeful, and I suffer many disappointments."

Electricity Is Life

Although Caroline believed that the prolonged galvanic treatments helped improve Carolyn's hearing, by and large electrotherapy was a marginal practice, existing on both ends of the therapeutic spectrum: in the clinical realm of hospitals specializing in nervous diseases, and in the thriving trade of electric medical appliances advertised by electric entrepreneurs.[28] In both cases, practitioners' claim of electricity's efficaciousness for particular diseases and disorders was met with skepticism by the larger medical profession, for extravagant claims were perceived as a slippery slope to outright quackery.[29] As historian Iwan Rhys Morus points out, "If electricity was the stuff of life, as many argued, then it seemed reasonable to suggest that a judicious dose of galvanic fluid to the diseased part of the body might bring about revitalization."[30] Offering the opportunity to intimately link electricity to the vital forces of life, electric entrepreneurs advertised a panoply of devices—belts, hairbrushes, collars, corsets, buttons, and clothes—and the many ways electricity could revitalize health. Endorsed as a universal cure-all, electrotherapy became, as many advertisements boasted, "the marvel of the century."[31]

The marvel began with experiments on frog legs. During the late eighteenth century, Italian physician Luigi Galvani discovered what he termed "animal electricity": that, when a frog's leg was connected to a metal circuit sparked by electricity, the leg's nerves and muscles twitched. Though Galvani's theory was partially incorrect (he attributed the twitches to "vital fluid" that generated electricity), later experiments built on his work, investigating how electric currents applied to body tissues can stimulate the contraction of different muscles, reinvigorate the nervous system, or, in the extreme, re-animate a corpse, as described in Mary Shelley's *Frankenstein*. The recognition of electricity's curative properties and its painless application—provided the practitioner insulated the patient properly be-

fore electrification—transformed it from a source of entertainment
to a profitable source of therapy.[32]

Aurists certainly embraced the option of galvanism as a treatment
for deafness that was diagnosed as paralysis of the auditory nerve.
If paralysis prevented sound vibrations from being transmitted
properly through the eardrums, then, they hypothesized, galvanism
could spark life into weakened nerves, thereby recharging the auditory system and curing deafness. While some physicians constructed
machines to safely apply electrical currents to the ear, as Angell did,
others criticized the profession's frenzy over an unconfirmed treatment method. Aurist John P. Pennefather, for instance, wrote in
his 1873 treatise *Deafness and Diseases of the Ear*: "I allude to this
vaunted remedy from the specious character it represents, and the
conquest frequently with which persons suffering from are tempted
to give it a trial, in many cases a prolonged one, to find themselves
in the end but disappointed dupes." He clarified that he was not de-
crying the valuable agency of medical galvanism nor contesting its
efficacy in treating auditory afflictions, but insisting that in cases of
"incurable deafness," electricity was useless, if not outright dangerous. Indeed, aurists used electricity as one of several therapies in a
regime for treating deafness but rarely employed it as a first-resort
treatment.[33]

By the late nineteenth century, medical electricity rapidly
emerged in the realm of consumer goods, as the advent of mail-
order catalogs offered consumers a range of electrotherapy products
for curing everything from cancer to headaches.[34] Medical electricians bemoaned the encroachment of capitalism into their domain.
As they struggled to preserve the legitimacy of their specialist niche
against professional skepticism, at the same time they also strove to
protect electrotherapy from the overwhelming intrusion of charlatanism. This was especially imperative as many of the fad products
blurred the line between what the medical profession accepted as a
"reputable device" and what was sold as a "consumer product," much
like Joseph Toynbee's design of the artificial eardrum. Likewise, as

historian Anna Wexler argues, the medical battery (a portable shock-producing device) was simultaneously considered a medical device even as identical quack versions were sold to consumers through newspaper advertisements using cure-all marketing language.[35]

The demarcations of legitimacy, however, probably made no difference to the consumer. They likely purchased electrotherapy products after being seduced by advertising and sales pitches of "paradisiacal vistas of health," where they could achieve a fantasy of a healthy and well-balanced bodily vitality.[36] Though aurists disparaged cure-all treatments for deafness by arguing that specific diagnoses required specific treatments, electrical entrepreneurs made no distinction between congenital and noncongenital cases in their appeal to sell as many machines as possible to consumers. Indeed, most commercial electrotherapy machines were promoted as universal cure-alls, with adjustments made for specific diseases and disorders. Some, however, did specialize in treating a single ailment.

One of the most illustrious machines marked as a deafness cure was developed and promoted in 1903 by the self-proclaimed "International Specialist" Dr. Guy Clifford Powell of Peoria, Illinois. Advertised as "Powell's Electro-Vibratory Cure for Deafness," the device came in two versions: "Model A" was a near-faithful reproduction of his patent, while "Model B" appeared later, in 1905, as a compact and portable version. The device worked by pumping air into the patient's ears through cotton-covered electrodes soaked in salt water, after which a jolt of electricity was generated by solenoid cells and transmitted into the ears. Two "Electro-Magneto Storage Cells" housed inside the top cover of the apparatus supplied the power. Of course, the consumer was cautioned to use only Powell's brand of battery, lest the apparatus get damaged.

"The Wonder of The Century," advertisements for the apparatus decreed, alongside a portrait of Powell in his mustachioed glory. This was a "Positive and Permanent Cure for Deafness" easily accomplished at home without the need for a physician, with "No Pain, No Detention from Work," though physicians wanting an advantage over their competitors could also purchase the machine. The accom-

FIGURE 3.2. The interior of Dr. Guy Clifford Powell's electrotherapy machine for treating deafness. Courtesy of Bob Greenspan, Collect Medical Antiques.

panying pamphlet outlined how Powell spent years researching how to harness the "wonderful Nature Forces" together to cure deafness, eventually succeeding such that all cases of deafness and head noises will "disappear like magic," no matter the cause or duration. Like sellers of patent medicine, Powell incorporated scientific details in his advertising as a way to validate the integrity of his product; his frequent reference to his patent, moreover, was also used as a way to legitimatize his commodity and expand its sales.[37] Initially priced at $100, Powell's apparatus was sold to consumers on both sides of the Atlantic who wrote to request a trial for home or office use. Powell then mailed the device once payment was received. He offered some clients hefty discounts to close or repeat a sale—in some instances, they only had to pay Powell $15!

While it is difficult to say how many consumers purchased Powell's device, it certainly was advertised enough to attract its detractors. Muckraker Samuel Hopkins Adams warned citizens not to purchase Powell's machine, claiming that it wasn't worth $100, $30, $25, or even 25 cents, except as a patent right that could be licensed and sold to other manufacturers. As typical of his investigative reports on patent medicine cure-alls, Adams wrote a scathing report of Powell's operation, claiming that despite Powell's insistence of professional legitimacy, he met every criterion that defined the quack industry: he made false claims, he made use of catchwords to enhance the product's scientific merits, he recruited religious sponsors to assert his credibility, he fluctuated widely with his price, and he based his business largely on correspondence, sending out identical form letters to prospective patients outlining similar treatments, no matter the diagnosis.[38]

Like Powell's device, other machines for deafness cures marketed by electrical entrepreneurs made their rounds on the advertising circuit, all making grandiose claims for scientific efficiency, including the Patented Rice Oto-Conussor, the Ear-O-Tone, and the Electro-Phonetic Treatment.[39] The "Magno-Electric Vibrator" invented by Thomas A. Edison Jr. conquered suffering and diseases using "a new force of science" for treating rheumatism, catarrh, consumption, paralysis, defective vision, and deafness. Likewise, the "Electricon" restored lost vitality and activity to the auditory nerves by direct application of "mild, continuous current of galvanic electricity," such that even "the deafest of the deaf are made to hear without medicine, without the surgeon's instruments, without trouble or inconvenience." As explained in a leaflet, the Electricon was not an artificial eardrum, patented nostrum, or prolonged medical treatment; what it actually is, they never tell us.[40] Then there was Dr. Samuel B. Smith's "Torpedo Electro-Magnetic Machines," which apparently won the Premium Medal for "Best Electro-Magnetic Machine" at the American Institute fair in New York in 1845. "This machine is called the Torpedo inasmuch as it gives a full shock, both from the Battery and from all parts of the Instrument itself, just like the Torpedo fish."[41] Customers could purchase their machine in a rosewood or zebrawood case. This electrotherapy treatment continued selling until the 1920s, promising that "even the born deaf are restored to hearing," a bold claim given that by that period, it was common-place knowledge that nothing could be done to correct congenital deafness.

Another practitioner who blurred the line between reputable service and charlatanism was George M. Branaman, who sold ear drops and an "Electro-Magnetic Head Cap" out of his Branaman Remedy Company in Kansas City, which was also advertised as Branaman's Medical Institute. One satisfied patient under his care, Virginia P. Dean of Montgomery, expressed her gratitude: "I am feeling alright myself and don't have any pain about my head or ears either thank God my hearing is just as good as it ever was and there is no more dryness in the Ears since using your treatment. . . . So you can put me down on your list as one more Patient cured."[42] As with other mail-order schemes, Branaman advertised widely, directing prospective customers to write to receive "symptom blanks," forms outlining health-related and personal questions to be assessed for a "personalized diagnosis." Such "mail-order doctors," writes Daniel J. Robinson, made use of new forms of marketing and opportunities for commercialization in the patent medicine trade. Elements of this trade—testimonial advertising, advice letters, sales letters, mailing lists, and letter brokerages—all "formed an integrated 'chain' of activities" that enabled the seller to hawk his product. Customers were certainly incited by Branaman's advertisements headlined with "FREE DEAFNESS CURE," promising to send two months' worth of medicine for free to anyone who wrote in. "No matter the responses to its symptom blanks," Robinson asserts, "the Branaman company invariably recommended the purchase of an eight-dollar electro-magnetic headcap to accompany the 'free' medicine it provided."[43]

"It is not surprising," expressed Arthur J. Cramp in *Hygeia*, the AMA's popular magazine for lay readers, "that phenomena so mysterious and even yet so little understood as electricity and magnetism should have been invoked both by ignorant enthusiasts and by venal quacks."[44] Electrical entrepreneurs may have exploited the public preoccupation with health and vitality, but as Takahiro Ueyama points out, people also believed diseases and disorders were caused by disturbances in the body's electrical equilibrium.[45] In such a privileged "neurasthenic nation," where the body functioned like an electric system, the busyness and encompassing realities of modern life took a toll on people's health.[46] The depletion of nervous energy then made people sick, weak, and disabled. And as electrification eclipsed cities and slowly illuminated American homes, electrotherapy was perceived as the obvious scientific solution for dealing with the anxieties of everyday life. For the deafened diagnosed as incurable, it was probably the only treatment worth trying.

Forces of Nature

Like shocks, which created an immediate sensation, light and vibration became part of a therapeutic performance. From the early 1890s, physicians increasingly used heliotherapy (sunlight) and phototherapy (artificial light) to treat a host of disorders, including tuberculosis, scrofula, and psoriasis.[47] Light, these physicians believed, was a natural force necessary for conquering every disorder, from serious diseases to superficial ailments. Following Danish physician Niels Ryberg Finsen's invention of phototherapy, light could even be harnessed in a hospital setting and when applied could aid in activating the body's "vital forces." As historian Tania Woloshyn explains, when light was directed onto a patient's skin, it was "perceived to stimulate the body's mechanisms, the tissues, organs and blood, and those internal processes were signaled by external signs occurring on the surface, most notably through the ongoing pigmentation of the skin [tanning]."[48] Some physicians claimed that direct application of ultraviolet rays followed by a regimen of electric vibration would enable electric waves to pierce the entire body and refresh each nerve and cell with electricity.[49]

Electric entrepreneurs adopted the principles of light therapy. In London, for instance, Martin Kroeger published several advertisements for his Kroeger Light Cure Institute, where certain types of middle-ear deafness could be cured. In *Surdus in Search of His Hearing* (1906), writer Evan Yellon documents his experience touring

Kroeger's institute and Kroeger's unique tripartite method of light, vibration, and electricity.[50] First, Yellon was taken into a dark room furnished with a couch and several apparatuses of "curious shapes" designed to pulsate powerful and concentrated rays of light. According to Kroeger, the machines have a threefold effect in treating deafness: "they first destroy any bacteria, or the cause of any growth, and secondly, they cause the blood to flow to the surface of the illuminated parts; and thirdly, they improve the quality of that blood by increasing the number of red pigments in it, a matter of very great importance when a malignant growth is to be resisted or an unhealthy condition resists."[51] The lamps gave out no heat, as Yellon discovered, and thus were presumably safe to use on a person's ear.

The tour continued to the Vibration Room, where at the center, a powerful electric motor was suspended in the middle of a massive brass frame. On the wall next to the frame was a switchboard. As Kroeger told Yellon, a patient was required to sit on a chair, upon which he would then connect a "queer little instrument" to the motor. The instrument was a vibrator with an ebony handle on one end with a wire connecting to the motor; at the other end was an oval pad fixed with a tiny engine that delivered an electric current when turned on.

As electrical entrepreneurs claimed, vibration was useful for treating cases of dry middle-ear catarrh, a type of deafness caused by the buildup of calcium in the small bones of the ear (ossicles). By supplying small amounts of current, vibration could break up the calcification and make the ossicles "movable" again, and thus enable sound waves to properly resonate in the auditory canal. The third and final part of Kroeger's method was in the Electricity Room. Yellon observed that the place was "literally bristle with wires, switches, and indicators," with wires connecting to a couch in the middle of the room. High-frequency currents of 200,000 volts were passed through the patient's body in order to strengthen and tone up the nervous system—and for deafness, to "strengthen the auditory nerve"![52]

Kroeger's method was hardly unique. Indeed, hospital patients in

Britain as well as in America were subjected to similar treatments. However, as Yellon points out, Kroeger's boastful advertisements and the secretiveness in which he cloaked the institute brought the legitimacy of his enterprise into question. After all, one distinguishing feature that separated orthodox practitioners from opportunistic charlatans was a general cultural agreement that "gentlemanly" physicians did not conceal their methods from their profession. An expanding popular press promoting electrical health goods further led the profession to deride electrical entrepreneurs as quacks for advertising swindles and useless gadgets to the public. Some medical professionals, however, did advertise, even collaborating with medical marketers, thus entangling dividing lines between the two groups.[53] Furthermore, as Anna Wexler argues, while one of the defining features of medical quackery was the secrecy of drug ingredients, "the same criterion did not hold true for electrotherapy devices: anyone . . . could open up the product and see how it worked." Overt commercialism then, became the mark of an electrotherapy charlatan.[54]

Kroeger eventually abandoned his institute and became one among the growing number of entrepreneurs marketing electrotherapy products for home use. During the turn of the century, high-frequency generator machines became trendy. These were machine that streamed purple light through glass tubes, that when applied as localized treatment (especially for skin conditions), could miraculously heal within a few treatments. The Toronto-based manufacturer of the "Branston Violet Ray High Frequency Generator," for instance, advertised to American consumers that it could "strengthen your nerves like magic." Safe for even the most delicate invalid or child, the device was promoted for at-home self-treatment, with a discount offered for first-time customers at the cost of $34.50, rather than the usual $40.00. The company asserted the generator could cure any ailment if used properly and frequently. Deafness in particular required extra patience and a longer trial period to obtain successful results; after a single session, users would notice that the roaring, ringing noises in the ear would have disappeared.[55]

Vibro-Devices

One brisk November day, I drove to the Science Museum of Minnesota's offsite warehouse in St. Paul to examine its collection of electrotherapy devices. I was particularly intrigued by a "Vibrometer" listed in the museum's database; since no photographs were appended, I wondered if it bore any resemblance to a device marketed by the Vibrophone Manufacturing Company of Brooklyn: a violin attached to a stethoscope and housed in a wooden box. The museum's Vibrometer was in a felt-lined wooden box, but instead of a violin, it was a banjo-shaped instrument with four strings stretched across its length, resting on a bridge. On each side of the bridge were two horizontal bars that, upon closer study, appeared to be connected to a yoke revolving in a shaft connected to a broken motor and belt. I deduced that the apparatus functioned by connecting rubber inserted into the ears to a missing part of the banjo that sent vibrations into the tympanum. Further research revealed that this apparatus was invented by Henry F. Garey, a Baltimore-based homeopathic physician, as a massage treatment for deafness. Garey first presented his creation at a meeting of the American Institute of Homeopathy in June 1892, and marketed it only for professional use, warning that to achieve unqualified success, patients were prohibited from using competing machines so as not to compromise the benefits of the Vibrometer.[56] For the brief time the instrument received attention, it appears aural specialists found it useful: one physician used the Vibrometer for 150 patients over the course of a year for patients with catarrhal deafness.[57] In only 2 percent of cases, however, was the Vibrometer successful in relieving deafness.

Of course, not all inventors marketed their vibro-devices directly to the medical profession. Some pleaded with the AMA for promotional assistance, as Joseph A. Danis of Niagara Falls did. After spending six months showing deaf people how to use his apparatus—a coiled spring with vibrations generated by contact with a phonoautograph record (though he does not elaborate on the pre-

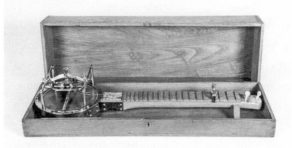

FIGURE 3.3. The Vibrometer was developed to supply an electric current to massage the middle ear to remove deposits and use vibration to restore the function of ossicles. Science Museum of Minnesota.

cise details of how the device worked, so as not to reveal trade secrets)—and contacting the Volta Bureau with notices that he was "curing hopeless cases of deafness," Danis hoped that the AMA would publicize and "enable the deaf throughout the world an opportunity" to try his novel creation. Unable to incur the costs of patenting his device, but certain of its ability to provide a "complete cure," Danis felt that such an instrument was crucial: "the mental attitude of the hopelessly deaf must be overcome and in order to do this my methods should be endorsed by the authorities they look up to."[58] Both the Volta Bureau and the AMA denied Danis their endorsement, with Cramp writing that it would be impossible for the organization to investigate every inventor's professed deafness cure that came their way. The AMA Bureau of Investigation, after all, answered nearly ten thousand inquiries annually from physicians and laity.

Handheld vibrating massagers were additionally advertised in catalogues and newspapers as inexpensive alternatives to high-frequency generator machines. The massaging system was particularly appealing for deaf customers, for as manufacturers repeatedly claimed, the vibration was a natural force that could "break up" deposits in the ear diagnosed to be the cause of hearing loss. Powell, for instance, offered a vibratory attachment for his apparatus: a patient would wear a stethoscope-like rubber tube to receive pulsating bursts of air pumped by a leather diaphragm; a sliding weight was attached to manually adjust the vibrating frequency.[59] Two years after introducing the Electro-Vibratory apparatus, Powell advertised a sleeker nonelectric version of his patented design, marketed as "Powell's Vibrator." This "stethoscope-without-its-chest-piece" apparently could cure deafness through sucking.[60] Users were directed to place the vibrator tubing into the ears and mouth, hold the ends in the ears, "then suck gently with the mouth for about a second, then stop for a second, and keep repeating this operation for about three minutes." Afterward, the user should repeat the letter B softly for thirty seconds, still holding the mouth over the end of the tube. Supposedly, this procedure would vibrate the ossicles, thereby delivering sound directly to the auditory nerve and "stimulating it to activity."

Some manufacturers even asserted that specific types of deafness could be cured only with vibration, as did T. H. Stilwell with his 1867 "Organic Patent Vibrator," which was marketed for "scrofulous deafness." Operating out of Toledo, E. D. Moon's "Ear-O-Tone" could improve 95 percent of long-standing cases, whether environmental or hereditary. Working by "immediately establishing better circulation within the ear," the Ear-O-Tone opened up clogged Eustachian tubes with vibration to "release" the ossicles to allow "them to again start carrying vibration back to the ear nerves and onto the brain." The device was priced at $25.00, and skeptical consumers were encouraged to take advantage of the company's ten-day free trial to give it an honest test.[61] Other manufacturers addressed the physical forces of nature that were harnessed in the motor: P. A. Grier of Cleveland marketed its noiseless, highly-polished aluminum "Royal" and "Rex" vibrators as "Nature's First Aid." And Dr. C. M. Jordan's fanciful "Aural Vibrator," a compact home instrument, operated by hand crank, was designed to stimulate the entire auditory mechanism—a procedure Jordan termed "aural kinetics"—which was apparently essential to perfect hearing.[62]

Some of the strangest and most successful vibro-devices of the early twentieth century were created by Charles Fensky. A law-

FIGURE 3.4. The Jordan Aural Vibrator in use. Warshaw Collection of Business Americana–Medicine, Archives Center, National Museum of American History, Smithsonian Institution.

yer based in St. Louis, Fensky filed for several patents for ear inserts that he asserted were based on extensive studies on the scientific principles of acoustics. In 1924, he filed a patent for a device he named "Audiophone," made of a hollow chamber and silver in an *L* shape. Available in five sizes, the device was practically "invisible" when worn, or at least that's what the advertisements said, though in reality one end stuck out of the ear canal, as observed in pamphlet photographs. At the Bernard Becker Medical Library at Washington University, I held a pair in my hands, struck by how large the devices

FIGURE 3.5. Advertisement showing Jordan Aural Vibrator in use (n.d.). Warshaw Collection of Business Americana–Medicine, Archives Center, National Museum of American History, Smithsonian Institution.

actually were, and how impossible it seemed to insert them into the ear and wear them comfortably. Though the patent specifications allude to the design's similarity to artificial eardrums, Fensky's creation was quite novel: inside the device was "a charge of radium so that diseases of the inner ear may be readily treated with radium."

Since its discovery by Marie Curie in 1898, radium entranced

FIGURE 3.6. Charles Fensky's Vibraphones (c. 1920s). Bernard Becker Medical Library, Washington University School of Medicine. Photograph by author.

Americans as a panacea that could cure everything ailing both the public and private body. As with electricity, patent medicine vendors exploited radium's glow, claiming that their products harnessed the element's energizing powers. While muddling scientific ideas about radium's benefits, these vendors persuaded thousands of Americans to purchase products infused with radium. Indeed, when it came to radium, explains Robert Wendell Holmes III, "Americans were ready to swallow anything, figuratively and literally."[63] Five years after filing the patent for his Audiophone, Fensky jumped on the "liquid sunshine" bandwagon, marking new ear inserts he named "Radium Ear." The primary difference between this and previous inventions, he claimed, was that the Radium Ear contained a "permanent radium deposit" called "Hearium," which was allegedly discovered by Abbott E. Kay of Glen Ellyn, Illinois. Yet Fensky appeared to have used the same ad format as he did for the audiophone but printed the Radium Ear upside down to differentiate the design. It is essentially the same product. Furthermore, Fensky gave no explanation for how radium could reduce hearing loss. It did no more for curing deafness than holding a luminous watch dial to the ear.[64]

Fensky later filed a patent for yet another modification of the Audiophone: a Vibraphone with grooves on the exterior shell and a small silver disc inside the chamber that vibrated when in use; it purported to "relieve hearing by utilizing and intensifying sound waves," though there is no indication that was possible. The "Fensky Quackery," as the AMA branded the scheme, sold Vibraphones for $15 a pair, though it likely cost one-third of the price to manufacture. Advertisements promised that the company would "refund one-half of the purchase price upon the return of a pair of them with the statement that no benefit [was] accrued." There's no evidence that this promise was honored.[65] Eventually, the Vibraphone Company was forced into an agreement with the Federal Trade Commission to eliminate all misleading claims in the advertisements, thereby joining the ranks of "quack schemes" eradicated from the marketplace.[66] Shortly afterward, Fensky sold his patent to the new owners of the Vibraphone Company. The company ceased to pay Fensky a royalty, and he no longer had anything to do with the device he created.[67]

Despite the AMA's best efforts to discredit instances of fraud, electrotherapy products harnessing natural forces of light, vibration, massage, and radium nevertheless made their manufacturers wealthy. As electrical entrepreneurs were aware, consumers were more likely to trust a product's benefits if it produced a sensation they could feel, see, and hear: spirals of sparks, hairs raising, the low hum of generators were often preferable to visiting a specialist and undertaking a prolonged course of treatment. And sometimes, if the electric sensation was felt strongly enough, it could create a placebo effect.

Pocket Battery

Not all consumers wanted, or could afford, the electrotherapy machines or handheld vibrating devices that glutted the medical marketplace. Some merely wanted the shock-producing benefits of electricity in a compact form: the medical battery, which produced low levels of direct and alternating current to deliver electrical stimulation to treat a host of diseases.[68] Also referred to as "family batteries," these devices could be purchased through mail-order catalogues or from a drugstore. These family batteries, explains historian Lisa Rosner, "encouraged the idea that anyone, no matter how limited his knowledge or training, could apply electricity, just as self-help medical books encouraged the idea that anyone could practice medicine."[69] Moreover, since the medical battery was also listed in reputable electrotherapy instrument catalogues and in physician's textbooks, and manufacturers marketed to both physicians and consumers, distinguishing whether a particular battery was quackery was difficult.[70]

One of the most notorious electric pocket batteries was "Actina," which was aggressively marketed as a "positive cure" for eye and ear diseases. It originated in Bristol, England. There, William Cotter Wilson opened a factory in the 1870s to make and sell "galvano-magnetic clothing." After working as an apprentice cabinetmaker and auctioneer, Wilson relocated west after realizing that the future resided in the remarkable benefits of electricity. Galvanism was quickly becoming medicine's most powerful remedy. Fifty girls were employed in the factory to make Wilson's "Magnetic Body Wear," sewing cotton and linen fabrics to magnetized steel plates attached to strips of copper and zinc. Far from being fashionable, this clothing line became part of a new trend of body-work galvanic products—though any electrician would have known Wilson's claims were technically nonsensical.

After immigrating to New York City in 1880, Wilson began to advertise his "Wilsonia" line of magnetic garments, promoting everything from "Lung Invigorators" to "In Soles" for aching feet.[71] "Take medicine and die," one advertising insert warned consumers; "Wear the 'Wilsonia' and live." Unsurprisingly, not long afterward, the company faced serious legal opposition. Robert K. Waits describes how Anthony Comstock, founder of the New York Society for the Suppression of Vice, investigated the integrity of the Wilsonia advertisements. By 1883, accusations of fraudulent testimonials and exaggerated advertisement claims, as well as libel suits, resulted in Wilson losing his Wilsonia Magnetic Company, leaving him in serious debt to multiple creditors. To avoid paying his debts, Wilson relocated to Philadelphia. In February 1886, he created a trademark for a new product: the Actina.

Described as a medicinal vaporizing device (that is, an inhaler), the device was a small steel vial, about three inches long with screw stoppers at both ends. Supposedly, one end of the vial cured eye ailments, the other ear troubles, and both worked simultaneously to cure a variety of illnesses. The principal medicinal ingredients were housed inside the cylinder: oil of mustard, oil of sassafras, belladonna extract, ether, amyl nitrate, and alkaloid atropine. Once the screw stopper was opened, the vapor would be released, and could "open deposits" to heal eye and ear diseases.

One of the earliest advertisements for the Actina appeared in an 1885 issue of *Frank Leslie's Popular Monthly*. Declaring the device as the "Wonder of the 19th Century" that harnessed the energizing forces of electricity, the advertisement assured prospective customers that the Actina was trademarked and patented, though the patent specifications themselves make no mention of the device's electric properties! Other advertisements stressed that 95 percent of all cases brought to the company's attention were the result of chronic catarrh of the throat and middle ear, thereby clogging the Eustachian tube and stopping the action of the "vibratory bones." "Nobody need to be Deaf where the Actina Pocket is Obtainable." One 1903 advertisement even boasted that Actina was a scientific cure of the highest order that had never failed to deliver.

Renaming the company "N.Y. & London Electric Association" and branding himself as "Professor Wilson" (though he lacked any claims to that title), Wilson relocated to Kansas City in 1890. He modified the device, lining its rims with zinc and copper, for supposedly the combination of the two metals caused a "bi-metallic" reaction to invoke galvanism. His concept was simple: "the human body is an electric battery, the upper half being positive and the lower half being negative . . . Electric Currents . . . keep the body in motion."[72] The new Actina was a marvelous "pocket battery," "electric battery," and "ozone battery" that served both young and old. It cured deafness and blindness and a host of other ailments, including hay fever, headaches, and neuralgia. Company pamphlets reassured customers that the product was not a nostrum, but a marvel of the century's new science, an infallible electric remedy that positively cures deafness.[73]

The N.Y. & London Electric Association prospered by the turn of the century, with offices throughout the Midwest to target unsophisticated potential customers. The Actina sold for $10, which, as Waits points out, was a week's pay for most men—as a point of comparison, men's suits were selling in department stores for $10–$15. Historian Carolyn Thomas de la Peña adds that body-worn electrical products or pocket batteries were often promoted with installment payment options that were attractive to working-class men who could not afford "proper" housing or food but had products like an electric belt.[74] For these men, the relatively inexpensive price of these devices and the appeal of home treatment likely guided their purchase, especially in cases where an illness had to be hidden to secure employment.

As with other mail-order products, interested customers wrote in for a trial, sent payment, and received the Actina through the post. Customers were also instructed to return their Actina every four months for "battery recharges" at a discounted cost of $1.00. The charges were nothing more than a reloading of the cylinders with Wilson's oil concoction, but they provided the company a steady stream of revenue. By 1896, frequent customer complaints drew the attention of ophthalmologists and the AMA. Declaring the Actina as "easily amongst the first mechanical fakes," the AMA estimated that over 110,000 were sold since Wilson established his N.Y. & London Electric Association.

The Actina was repeatedly cited in congressional hearings leading to the passage of the Pure Food and Drug Act of 1906 as an example of quackery at its most notorious. Federal statutes to protect the integrity of the food supply and drug ingredients first began in 1850, when a law was passed to clarify the classification of tea. By 1906, 190 measures were presented in Congress to protect consumers from adulterated food and drugs; only 8 became law.[75] Of these bills, 141 were rejected outright, their opponents arguing that it was unconstitutional for the federal government to regulate state power. Capitalism also interfered in politics, as it often tends to do: an ideal pure food and drug measure would require criminalization of fraud or medical quackery, which would run contrary to some congressmen's business interests in the food and drug industry. Food manufacturers, for instance, used spices and other additives in their canned goods to mask the taste of expired meat. Meanwhile, patent medicine vendors included large quantities of morphine, opium, and cocaine in their drugs to increase sales. With the work of Harvey Wiley, head of the Department of Agriculture's Bureau of Chemistry, who organized a "poison squad" to test how food preservatives affected health, and the muckrakers who ruthlessly exposed corruption and medical fraud, the crusade for legislation was in full force during the 1900s.[76] Upton Sinclair's *The Jungle*, an exposé on unsanitary processed foods and working-class poverty, and Samuel Hopkins Adams's "The Great American Fraud" series especially commanded public attention to the issue. After years of efforts by various individuals and organizations, President Theodore Roosevelt signed the Pure Food and Drug Act into law on 30 June 1906.

Nicknamed the Wiley Act due to the tireless efforts of Harvey Wiley to improve food manufacturing, the act gave regulators unprecedented authority over interstate trade in food and drug products, including requiring all drug makers to disclose ingredients on product labels. Administered by the Bureau of Chemistry, administration of the law mostly focused on premarket approval of foods and chemical additives until 1912, when the bureau shifted to target the proliferating patent medicine industry and advertising frauds who heavily relied on mail-order sales. Effective enforcement, however, was challenging, especially when suspects simply relocated to another state to escape capture.

After Adams's exposé on the Actina's supposed electrotherapy benefits, Wilson renamed his firm the Actina Appliance Company of Kansas City, Mo., and continued to advertise his device, this time without any claims to electrotherapy. Advertisements in England, however, continued to reference the Actina's "magnetic" and "electric" features. Nevertheless, numerous fraud complaints were leveled against Wilson. On 24 May 1914, Judge W. H. Lamar recommended that the postmaster-general place a fraud order against Wilson's firm to prevent mail-order service. Prompted by further investigations by the AMA's Propaganda Department, a year later Congress launched a case against the Actina Appliance Company. During the three-day hearings in Washington, D.C., testimony confirmed that the Actina had no electrical properties. Wilson never appeared for his trial, but his company's president, John Foran, admitted that the device was not electric and that neither he nor Wilson possessed the medical credentials they claimed in their advertisements.[77]

After a twenty-year success, the Actina was deemed a worthless quack product by the federal government. It eventually disappeared from advertised sections of periodicals after 1916.

Other advertised pocket batteries offered customers alternatives for electrical deafness cures. For fifty cents—a fraction of the cost of the Actina—Brewster's Medicated Electricity promised to "instantly relieve, the cure is truly magical." Used properly, the "medicated vapor from the battery," the company claimed, will produce a permanent cure in one or two weeks. No recharging necessary, no extra costs—it's always ready to use and convenient to carry. The Davis Electric Battery was advertised as the "new way to health" for general or constitutional treatment, with treatments taking only five or fifteen minutes a day. To treat deafness, customers were instructed to use the nickel electrode attachment with a wet cloth and apply the electrode on and under the ears with their mouths open;

for added benefits, they should place the electrode on the chin with teeth closed. "There are no cells—no wires to rust or corrode—nothing to get out of order, and with proper care it will last a lifetime." The Davis Electric Battery was so simple that any child could operate it: "You simply unscrew the electric light bulb from its socket and insert our attachment plug in its place, connect wires to the battery and it is ready for use."[78]

Electric Impulses

As outlined by the 1906 Pure Food and Drug Act, a telltale feature of deceitful trade practice was the dabbling in "side-lines": commodities that exploited consumers, sold as add-on worthless "courses of treatment" or devices. These products enabled manufacturers to sell more products and increase their sales, kind of like how television infomercials contain "BUT WAIT—THERE'S MORE" to convince the customer to take advantage of a one-time deal and get more goods at low prices. George Way, for instance, encouraged customers purchasing his "Way's Artificial Eardrum" to add on the "Blowena," a small plastic device for treating catarrh, hay fever, head colds, the flu, or any irritation of the nose and throat that were symptomatic of deafness. Electrical entrepreneurs also sold side-lines, as Powell did with his vibratory stethoscope. Consumers purchasing a handheld massage device could also buy separate attachments that would give them options for treating diseases and disorders.

While the 1906 act marked the start of federal regulation against deceptive practices and harmful substances, medical devices were not included in the act. It became quickly apparent to the FDA that patent medicine manufacturers and electrical entrepreneurs could exploit this loophole by marketing products or side-line cures to escape regulatory inspection, like the "Lash-Lure," an eyelash dye that injured several women and resulted in one confirmed case of blindness.[79] While several companies supplied medical instruments and equipment to physicians and hospitals, most commercial medical devices posed little danger to consumers. By 1917, the business of medical devices flourished, in part due to the limitations of the 1906 act, and in part due to the electrification of goods. Electrical contrivances such as height-stretching machines and color-therapy apparatuses were advertised for their ability to treat disease or improve health—and most, if not all, were considered fraudulent by federal regulators and physicians.[80]

Hearing aid manufactures also positioned themselves within the gray area of FDA jurisdiction. "There are on the market," wrote Arthur J. Cramp in the third edition of *Nostrums and Quackery*, "a number of meritorious aids to hearing." Constructed on the principle of a small portable telephone with a microphone attachment, most electrical hearing aids could mitigate various kinds of non-congenital deafness, but no user should expect to be "restored to normal." However, Cramp lamented, it was "unfortunate that some of these electric aids, while valuable in themselves, have been sold under exaggerated claims."[81] Preposterous falsehoods, misleading statements, and outrageous expectations of "cure" were, again, problematic advertising tactics directed towards selling the incurable deaf yet another "hope" for a cure. As the *Silent Worker* editorialized, while hearing aids assuredly had some benefit, they were not meant to be "miracle" devices, nor they could improve the lives of all deaf persons—even if advertisers insisted otherwise.[82]

Still, some people worried that the AMA's stance on hearing aids was unnecessarily prejudicial and that Cramp's persistent crackdown on mechanical quackery unfairly dismissed useful commercial hearing aids in favor of advocating specialist physician care. Olive A. Whildin, the supervisor of the Baltimore Public Schools' deaf and hard of hearing classes, for instance, notified concerned parents that although there were several "splendid instruments" marketed for amplifying sound, they were at best, only a "crutch for the ears."[83] As letters in the AMA archives reveal, deaf customers were often frustrated by the lack of effective treatment by ear specialists. Oscar F. Swenson of Poughkeepsie knew that most deafness devices were fake, but after several failed specialist treatments, he wrote to Cramp inquiring whether the Globe Ear-Phone was suitable for his deafness.[84] Elise D. Nelson, on the other hand, wrote to the AMA disagreeing with Cramp's stance on hearing aids: "Earphones are a definite aid to those who can use them, and it is my opinion that a deafened person should use medical treatments, lip reading and an earphone, all three to help him be as normal as possible."[85] Organizations such as the New York League for the Hard of Hearing also worked tirelessly to reconfigure public perceptions of hearing aids and ensure that deaf consumers were well-educated on the fact that hearing aids worked to amplify sound, not cure hearing loss.

The technical features of hearing aids had significantly improved since Miller Reese Hutchison introduced his Acousticon. Carbon electric hearing aids frequently had problems of microphone distortion and filtering external noise. Later design improvements included an amplifier (a booster) that allowed for indirect amplification of the electric signal to make sounds clearer.[86] However, these aids did not provide sufficient gain for users with more than moderate hearing loss, but at the very least they were practical and resilient to damage, and the internal noise was bearable; it could still be taxing for a user to wear this device all day, especially with the added weight of battery packs. The introduction of vacuum tubes in the 1920s enabled improved amplification of electric signals, leading to greater acoustic gain. This design was a better solution than increasing battery voltage or adding more microphones, as was featured in some carbon models.[87] Within ten years, nearly all hearing aids available on the market were vacuum tube models.

During the first three decades of the twentieth century, most pioneering hearing aid firms sold side-line electric devices as "cures" for deafness at discounted cost with the purchase of an aid. Such companies included the Hutchison Acoustic Company, Mears Radio-Hearing Device Corp. (est. 1904), the Globe Ear-Phone Company (est. 1908), and the Gem Phone of New York City (est. 1912). Much to Cramp's disdain, these electrotherapy gadgets targeted customers whose deafness was not, or indeed could not be, improved with a hearing aid. Though he admired the technological capabilities of hearing aids, he regularly decried companies' misleading advertising and their promotion of side-line electrotherapy devices.[88] All side-line treatments, he asserted, were, "without exception, humbugs."[89]

Hutchison, for example, initially marketed his new "Akou-Massage" directly to aurists to treat inflammatory deposits in the ear that cause middle-ear and catarrhal deafness. The device was priced at $60.00 (approximately $1,600 in 2016 dollars) and came supplied with a dry cell battery.[90] Under the re-formed Acoustic Company, Hutchison rebranded the instrument as the "Massacon" and promoted it to both medical professionals and the deafened public—though it was unaffordable for the average 1910s American worker whose salary was between $200 and $400 a year. Promotional literature stressed the medical and scientific merits of the device, quoting from one of Hutchison's London lectures: "It is a remarkable, although sad fact that, while we have made progress in almost every other line of electrical researches, [deafness] has been completely neglected; yet I know of none where the need is more urgent or the work more baffling."[91] With the Massacon, total or partial deafness "need no longer be a hopeless affliction." Hutchison's profitable Acousticon may have been invented as an artificial aid to hearing, but the Massacon was a cure for deafness.

Other hearing aid companies followed suit. Potential customers interested in the Gem Phone Company's commendable hearing aid were encouraged to additionally purchase the "Auto Ear Massage" at discounted cost. For $10.00 and a free trial, customers were told that the vibration would stop common ear pain and cure ear diseases by improving blood circulation.[92] Letters sent to the AMA affirm that even as physicians recommended particular hearing aid models that would best serve their patients—the Gem Phone, for instance, being a popular one—the appeal of a permanent cure as advertised by hearing aid companies encouraged patients to explore other avenues for electrical treatment.[93] Some Gem Phone users trying out the device under a free trial agreement were cautious about the benefits they received, and wondered whether their hearing would further improve with the company's vibratory treatment. A reader of *Life & Health* magazine, for instance, testified that the Gem Phone "helped me to hear better after I had used it a while," but was hesitant in purchasing because it was too expensive and she was unsure about the long-term benefits. The Auto Ear Massage, however, was cheaper so she debated whether to exchange the hearing aid for the vibrator.[94] She apparently decided to keep the Gem Phone.

Chicago-based Charles Endorf Jr.'s short-lived Telonor Corporation conveyed messages similar to the Gem Phone's. Their primary hearing aid, the Telonor Ear Phone with "Sensitone," a cylindrical pocket volume control device for adjusting sound quality, was promoted as a free bonus to customers who purchased the "Telonor Pneumatic Massage" and the "Telonor Vibrating System"—devices that, when used together, provided an effective means for treating deafness.[95] A hearing aid could only go so far, the company's advertisements proclaimed, for unlike those who were born deaf and have never heard sounds, the deafened "cannot become resigned to their fate. They are forever struggling and straining to re-enter this enchanted world of sound." The electrotherapy treatments offered the hopeful a chance at a permanent cure: hearing aids were a means of last resort, to be used only if all other treatments failed.

Even the second-oldest manufacturer of hearing aids promoted side-line electrical products. The Mears Ear Phone Company of New York City was established in 1904 by Willard S. Mears, who previously worked for Hutchison's company. After launching the Mears Ear Phone, the company began marketing a device called "Aurasage"—sometimes named "Airosage"—and Mears Ear Oil, claiming that when used together with their hearing aid, the "ease with which it improved the hearing of people who thought they were beyond help seems nothing short of miraculous!"[96] The Mears Ear Phone helped users hear instantly, but the Aurasage gradually eliminated the head noises and congestion largely responsible for deafness. Despite derision from the medical community, the Mears Ear Phone Company continued selling the Aurasage until the 1930s.[97] The company's emphasis on its "cures" as much as on its electric hearing aids eventually came under fire from the Federal Trade Commission (FTC). In 1914, partly in response to the limitations set by the 1906 Pure Food and Drug Act, the FTC was set up as a task force to regulate unfair business competition in commerce. On 2 February 1934, it sent out a press release prohibiting the advertisement and sale of the Aurasage/Airosage on grounds that the treatment was unscientific and dangerous.[98]

Around the time the press release was circulated, physicians, policymakers, and federal representatives were pressuring Congress to amend the 1906 act to extend its prohibitions to adulteration, drug misbranding, and fraudulent medical devices. The initial suggestion to expand the act's definition of drugs to include devices made little sense. Indeed, when the 1906 act was first drafted, medical devices were not as prominent on the market as they would be a few decades later; further, policymakers argued that national regulation of the medical device industry was required, especially for protecting consumers against fraudulent or dangerous home health products.[99] After five years of debate, a bill was passed by Congress and signed by President Franklin Delano Roosevelt on 25 June 1938. The new Food, Drug, and Cosmetic Act brought medical devices under FDA authority and equated their regulation to that of drugs in order to reduce or eliminate sales of fraudulent goods.

As historian Takahiro Ueyama asserts, "For consumers, buying and enjoying goods became a capitalist form of self-definition."[100] For deaf persons wanting to restore their hearing and disappointed with traditional medical therapeutics, the range of electrotherapy gadgets and machines presented a modern solution for asserting their needs for a permanent cure. From electric belts targeting neurasthenic symptoms to acoustic vibrators sold as side-line additions to hearing aids, these therapeutic commodities became novel reflections of how modernity and progress can electrify cultural expectations of normalcy. Deaf persons were encouraged by their physicians, family, friends—and, even by their own selves—to seek out any and all possible solutions to correct their hearing. Even as the sales of medical devices declined following the 1938 act and the FTC crackdown on mechanical quackery, these ideas nevertheless still resonated.

4

Fanciful Fads

Perhaps this "cure" is not as yet too untried to be considered reliable. But
at least it indicates a commendable effort at keeping up with the times
by seeking to apply development in one field of science for the good of
another.

Editorial, *Stanberry Headlight* (1927)[1]

It is easy to say, "Oh, there are worse things than deafness. After all, nobody dies of deafness." Let me tell you that people do die of deafness—
a hundred little deaths a day.

Frances Warfield (1952)[2]

Supposedly, during the First World War, a man "suffering with hysterical deafness went up in an airplane, received the shock of his
life, and came down quite able to hear." Likely this was thirty-one-
year-old serviceman August S. Harris, who lost his hearing from
exposure after he slept without shelter in a Georgia camp in 1917.
Declared "hopelessly deaf," Harris decided to experiment with airplane diving, having heard rumors that shifts in altitudinal pressure
was helpful for reducing swelling in the ears and "popping" things
into place to restore hearing. In September 1922, after consulting
with local officers from the American Legion, Harris arranged with
pilot Jimmy Curran to fly up in a plane. Several hundred spectators
gathered one afternoon in Ashburn Flying Field to witness Harris's
attempt at the "flying cure." An hour and fifteen minutes later, having
risen to an altitude of fifteen thousand feet, Harris and Curran re-

turned to the ground. Prior to the experiment, Harris could barely hear a ticking watch; after, he could hear people speaking to him, including his mother, who he telephoned upon landing.[3]

The national press sensationalized Harris's story and those of others who followed suit. Front pages across America covered deaf individuals being taken up for a series of nose dives, loops, spins, and barrel rolls as part of the flying cure. Multiple reports testified that once the plane landed, these "patients" maintained they could hear better; others claimed to completely regain their lost hearing. Some took rides up to ten thousand feet, even taking parachute jumps at the recommendation of ear specialists who theorized that changes in air pressure could stabilize the Eustachian tubes. Most physicians, however, were baffled as to why the flying cure seemed to work for individuals who were diagnosed with permanent hearing loss. Nor could they explain how the "treatment" even worked: was it a specific altitude? The "popping" of the ears? Were stunts necessary or just for show? Were the benefits permanent and deafness kept at bay?

Though quantifying hearing gain or improvement was difficult without proper audiometric data, and despite the fact that the accounts of "successful" rides were anecdotal, tales of the flying cure made for exciting news stories well into the 1930s. Whether the reports were exaggerated or not, they captured the attention of deaf people across America: the more publicity the flying cure received, the more people wanted to experiment with it.

Two years after Harris's experiment, on 12 December 1924, twenty-two-year-old Gwendolin Caswell, who had lost her hearing at age three after an illness, went up in an airplane outside of Chicago. While in the air, she could hear the airplane motor and pick up snippets of conversation with the pilot, but upon landing she realized that the benefits did not last. In Decatur, Illinois, Mille Lucinda, a vaudeville dancer who had been completely deaf for six years, tried the cure. On the west coast, Oakland pianist Emma Menz went up nine thousand feet in the hopes of resuming a musical career interrupted by her deafness. She "found a distinct improvement but a day later the improvement had completely disappeared," so she booked successive flights to lengthen the time before a relapse.[4] Postmaster Ray Little took a flight from Gettysburg to Baltimore and back, with a nosedive in between. As the velocity of the plane increased during its dive, Little felt a pain in his eardrum and a cracking noise in his ears, but no hearing returned. In Hutchinson, Kansas, Harry R. Atkinson sent his seventeen-year-old congenitally deaf son up on Ralph Heimer's new Spartan plane. Heimer flew nine thousand feet in thirty-five minutes, then climbed to ten thousand, after which he dove at 220 miles/hour, zooming to inflict quick pressure to the boy's eardrums. After an hour, they descended and the boy complained of pain behind his ears—encouraging news for his father, who then planned further flights to "help the boy."[5] Following an ear specialist's advice, six-year-old Barbara Michalovic's parents sent their daughter for the cure; the roar of the small engine could barely drown out the girl's hysterical screams as the plane proceeded in a steep descending spiral.[6] Even a dog was taken up: albino collie Bob White, the seven-month-old grandson of Rob Roy, the celebrated White House dog.[7]

Physicians also hopped on the flying bandwagon, recommending stunt flying for patients who had failed to respond to conventional treatments. A New York City doctor ordered a seventeen-year-old boy to take several flights as part of his ongoing treatment for hearing loss. Encouraged by Dr. Thomas F. McGuire, Miss Vera Price also took stunt flying.[8] So too did Mrs. William Schaffer of Brooklyn, whose ten-year-old son Julius was born deaf, blind, and mute; though the boy's hearing and sight were surgically restored by Dr. Adolf Lorenze, the mother hoped the flight would shock her son's system and give him speech. News reports on Mrs. Schaffer's attempts eventually drew the attention of the Brooklyn Society for the Prevention of Cruelty to Children, leading them to interfere and forbid further flights for young Julius.[9]

Plenty of commentators expressed their thoughts on this new fad. The *Wilmington News-Journal* editorialized in 1925 that despite anecdotal accounts, it did not appear that the cure would benefit all types of deafness, if at all. "It does seem a bit like the method in the old Chinese story," the reporter continued, "of burning down the house to roast the pig instead of using a small fire especially adapted to meat-roasting."[10] Furthermore, why go through all the trouble and expense of flying up to the clouds when the same thing—albeit less thrilling—could be achieved in an airtight room controlled by scientific and medical experts? The publicity bureau of the Indiana State Medical Association agreed. In an October 1925 bulletin, the association warned "too extravagant claims should not make [sic] for airplane rides as a cure for real deafness."[11]

Especially after all the deaths from this flying cure.

In September 1925, twelve-year-old Clifford Davis was killed when Captain Walter L. Smith, a private flying instructor, crashed his plane from more than ten thousand feet into Highland Park cemetery just outside Cleveland. Davis, who was undertaking the flying cure, died instantly as the crash was witnessed by his parents and several golf players. Two months later in Independence, Kansas, twenty-five-year-old Paul Gibson, who was deaf-mute since birth, and his pilot Harold H. Caulkins died when the wing of their airplane broke off and crashed five thousand feed to the ground. "Eight Killed in Sunday Series of Air Mishaps," the *Miami Daily News* front page reported on 30 April 1928, including three people who lost their lives near Eureka, California, in efforts to cure the deafness of composer William Spaletta. In Springfield, Massachusetts, an ill-fated loop broke the left wing from an airplane two thousand feet in the air, crashing the plane near Randall Field Airport. The pilot, a passenger, and six-year-old Luke Briotta were killed instantly. The pilot had been warned not to do any stunts, but the boy's father insisted in the hopes of restoring his son's hearing.[12]

The most sensational story of airplane diving occurred on 24 February 1930. Welterweight boxer Fred "Dummy" Mahan, deaf-mute since he was eight months old, decided to take a parachute jump after physicians told him it could restore his hearing. Mahan was well-known on the boxing circuit, having fought welterweight champion Jackie Fields, Young Corbett, and Myer Grace of Chicago and holding a fight record of sixty-four knockouts in one hundred matches. Though he previously arranged for an airplane dive in Los Angeles, the attempt was unsuccessful. Over the protests of his manager Fred Winsor, Mahan arranged for another jump from a plane piloted by Berkeley aviator Colonel Harry Abbot. At five thousand feet, Mahan leaped, but he pulled the ripcord too soon and the parachute failed to open. He plunged to his death in front of a crowd of several thousand persons.[13]

The increasing number of fatalities became concerning enough that aviation experts released statements to correct public misconceptions about the flying cure. Speaking to the Associated Press, Lieut. Col. Levy M. Hathaway, the chief medical officer of the Army Air Corps, declared that flying was an incredibly harmful endeavor as a "cure": forget the risk of plane crashes—engine noises and rapidly changing atmospheric pressure could burst weakened eardrums and permanently damage dull hearing. Furthermore, Hathaway explained, there was no foundation to support the idea that airplane flights could cure deafness; in fact, defective hearing was a common occupational *hazard* among aviators. These anecdotal reports were nothing but sensationalized fad stories.

Members of the medical profession affirmed Hathaway's statement, as did officials at schools for the deaf. Aviation stunts posed no more value as a cure for children with hearing or speech defects than "swallowing a decoction from a witch's cauldron brewed at midnight." This was simply another thrill serving to mislead "unfortunate sufferers by raising false hopes."[14] After two deaf boys, one in Ohio and the other in Iowa, lost their lives from an airplane dive, the *California News* periodical for deaf readers condemned the practice stating that the "risk hardly justifies the experiment."[15] This did not discourage deaf persons like Mrs. J. A. Deegan of Yonkers, New York, who wrote to Arthur J. Cramp in 1931 asking if airplane flights could cure deafness. Cramp replied, "Of course in cases of deafness of a purely psychological character airplane flights may seem to have curative results. So may carrying the left hind foot of a rabbit that has been caught in the churchyard in the dark of the moon!"[16]

None of these cautionary statements did much to dispel the flying cure fad. "Airplane treatment for deafness," reported *The Brooklyn Daily Eagle* in 1928, "which threatened to become almost as much of a fad as reducing or golf, is due for a tailspin, if the doctors have anything to say about it."[17] At Mitchell Field in 1932, the Association of Physicians of Long Island gathered for their monthly meeting, with a special symposium dedicated to aviation medicine. Majors from the Army Air Corps agreed that stunt flying had no value as a medical corrective for deafness, and in fact, did more harm than good. Dr. Louis H. Bauer of the United States Department of Commerce ridiculed airplane diving, arguing that hysterical deafness caused by emotional strain was probably the only type of hearing loss that might benefit from the treatment. Or, as Dr. George B. McAuliffe, professor of otology at Cornell University Medical School, observed, perhaps what was "cured" was not deafness per se, but a form of "shell shock."[18] Furthermore, Bauer noted that there was only one certain case in which a plane dive made a change in the condition of a deaf man: "It killed him."[19]

Healing fads, historian Claire Badaracco asserts, are rooted "in the deeper levels of our cultural heritage" and in the business of advertising hope.[20] They thrive on spectacle and fascination with the miraculous and the magical. They are temporary bursts of enthusiasm, but they linger stubbornly in the public imagination, fueled by the media's obsession with sensational stories and charismatic celebrities with their extraordinary treatments.[21] And fads flourish on an ailing person's optimistic desire for a cure, manifesting in different ways, but always rooted in the belief that a cure, no matter how implausible or dangerous, is important enough to warrant acceptance. Like the airplane cure craze, many fads soar as high as possible, only to swoop down in a series of loops, rolls, flips, and dives, before descending to the ground or crashing in a fiery end.

Medical fads often emerge in response to public eagerness for scientific innovations, or in response to a newly discovered health problem, much like the way fatness and posture became pressing health issues.[22] Part of these responses were derived from public frustrations with the discrepancy between the authority of physicians and their curative capabilities, which could aggravate populist resentments and lead to new health movements—like chiropractic or the water cure.[23] As fads and their fancies were disseminated through enthusiastic publicity and (anecdotal) claims of simplicity and safety, they often tended to be framed as "anti-intellectual" or as a rejection of scientific frameworks. As Holly Folk argues, however, fads, and broader populist health movements, are rather *counterintellectual*, "pursuing alternatives to mainstream knowledge, which they see as tainted by plutocrats and aristocrats."[24] This led to early twentieth-century American health becoming a "patchwork of experiences that differed according to race, geographic location, culture, and social class," as well as access to care. For some people, fads were more accessible and affordable than long-term care under a physician, with the added bonus that they could deliver a quicker cure. Though fads like John R. Brinkley's goat gland transplantation for impotence attracted national attention for its fraudulent claims, not all fads ended up being frauds.[25]

More commonly, however, fads gained prominence precisely *because* they were promoted as being effective. As historian John Burnham notes, the media's perpetuation of fads resides "not only in quests for cures but also in a continuing series of medical, public, and political concerns about diseases."[26] The emergence of "safe" surgery in the 1880s, for instance, transformed cultural expectations of the surgeon and surgical therapy, as an increasing number of conditions—such as appendicitis—became redefined as an exclusively surgical disease treatable by new procedures or operations.[27] Successful procedures were thus magnified by popular media, convincing people that surgery could be the only avenue for successful treatment. This was especially the case for progressive diseases, where fads were likely to emerge. When a treatment appears for chronic or acute conditions—like multiple sclerosis, mental disorders, or deafness—the possibility of a cure creates enthusiasm in both patients and physicians, spreading rapidly and becoming popular even if there is little clinical consensus to support its efficacy.[28]

Indeed, as George E. Shambaugh, one of the leading otologists of the early twentieth century—whose namesake son would prove equally famous—said, "Every field of medicine has its fads which for longer or shorter periods may occupy the imagination and satisfy the credulity of the patient as well as physician."[29] While health fads are characterized by a lack of scientific justification and the presence of interest groups who might profit economically, surgical fads tend to have different attributes.[30] They are usually procedures that have undergone numerous clinical applications, been published in learned journals, and been supported by followers who then establish new institutions to carry on the work. Surgical fads also tend to be routine procedures that become criticized for overuse or mismanagement.

Tonsillectomy for children, for instance, was a routine procedure at the start of the twentieth century. By the late 1980s, changes in medical technologies and expectations about the patient's role in health care led physicians and politicians to condemn the operation as a frivolous "wicked operation," a "dangerous fad" focused on aesthetic rather than medical value.[31] So too was bilateral ovariotomy, or Battey's operation (named after accomplished surgeon Robert Battey), which was devised to surgically produce menopause to alleviate a woman's gynecological pain. It came to be dubbed "castration," "unsexing," or "spaying." Moreover, the operation was criticized for being performed too frequently, as surgeons were flagrantly abusing the procedure.[32] Even the radical mastectomy, first performed by William Halstead in 1894, became the standard of care for breast cancer despite causing extreme distress for patients and despite physicians doubting its efficacy, arguing it was unnecessary in most cases.[33]

Just as the flying cure fad obtained legitimacy with physicians' public support, surgical and medical fads also were popularized by the charisma of their creators. They were media moguls who used their power and networks to bolster their procedures and scorned the medical establishment for restricting the American patient's need for therapeutic miracles. They were superb salesmen who were discredited as charlatans, even as they grounded their procedures in scientific medicine and amassed vast fortunes. And they appealed to the most basic and emotional source of vulnerability in their patients, the need for enterprising medical care rooted in optimism—no matter how far-fetched and no matter that medical authorities declared it more fanciful than factual. Some fads, then, managed to flourish during the twentieth century, despite good reason for skepticism.

Manipulative Magnetism

Perhaps no deafness cure fad was more egregious than the one concocted by Daniel David Palmer. Born in 1845 to shoemaker and grocer Thomas Palmer and Catherine McVay in Port Perry, Ontario, and the eldest of six children, Palmer was expected to shoulder the family's financial responsibility. At age eleven, he discontinued full-time schooling to work after his father's business failed. He eventually resumed his studies, but was interrupted again when Thomas decided to move the family to Iowa in 1856; Palmer and his brother Thomas J. decided to remain in Ontario. The American Civil War, however, overran the Canadian labor market with men fleeing from the U.S. draft, making work scarce and wages low. By 2 April 1865, the brothers packed up their belongings and headed south to rejoin the family. They walked eighteen miles to Whitby, then paid passage to journey on the Mississippi River, receiving permission from a military commander to hitch with his troops to Davenport, Iowa.

Eventually settling in New Boston, Illinois, Palmer spent the next twenty years teaching, raising bees, and selling raspberries in towns alongside the river before relocating back to Iowa to establish a grocery business. In 1884, he found himself a new vocation after meeting Paul Caster, a magnetic healer who toured America's heartlands. A popular form of healing after the Civil War, magnetism built on the idea that bodies are surrounded by invisible magnetic energy that could be manipulated by a skilled healer to cure most diseases.[34] Some practitioners, like Caster, went so far as to claim that they possessed "personal magnetism" radiating through their hands that enabled them to maneuver damaged magnetic energies and thereby restore the body to health. Guided by his mentor, in September 1886, Palmer opened his first magnetic healing office in Burlington, Iowa, moving his practice to Davenport a year later and advertising himself as a "Vital Healer" who "Cures Without Medicine." Though medical magnetism had passed its heyday by the time Palmer opened shop, upwards of a hundred patients a day packed his offices.[35]

A vocal opponent of regular medicine, including vaccinations, medical drugs, and vivisection, Palmer merged magnetism with his interests in phrenology and spiritualism, placing his life-renewing hands on painful areas of the body to influence nerves and the circulation of bodily fluids. He theorized that altered nerve flow was the cause of all disease and attempted to outline a relationship between the physiological and physical body.[36] Magnetism enabled him to restore the nerves to their positive alignment and thus heal all those who arrived in this office. His business grew tremendously: the original three office rooms were expanded to forty-two, occupying the entire fourth floor of a building. Treatment was $10 for the first week and $5 for each subsequent week—except for lupus, cancer, tumors, and other special cases that required more intensive care, including room and board.

"I do not claim to cure all diseases," Palmer stated, "but I now treat and cure many diseases which I had not thought of doing five years ago." His magnetism cured everything from rheumatism to skin disorders to cancer to deafness. One testimonial from Mrs. Joseph Rosette, who suffered from an aural inflammation resulting in gradual hearing loss, praised Palmer's approach: "I was treated by six doctors, but they did me no good; I tried bottle after bottle of drug store medicine and different kinds of salve, but all were useless. . . . My restoration is due to Dr. Palmer's magnetic treatment, for he gave me no medicines, inside or out, and only used his hands."[37]

Palmer's method of drugless healing, however, drew criticism from the local *Davenport Leader*, which labeled him a "crank on magnetism, [who] has a crazy notion that he can cure the sick and crippled by his magnetic hands. His victims are the weak-minded, ignorant and superstitious, those foolish people who have been sick for years and have become tired of the regular physician and want health by a short-cut method."[38] He was a mere mountebank, a huckster who appealed to a class of people believing in "spooks and other forms of occult things."

Palmer's magnetism was transformed on 18 September 1895 following an encounter with Harvey Lillard, an African-American janitor who worked in Palmer's building. Lillard had been deaf for nearly seventeen years and unable to hear "the racket of a wagon on the street or the ticking of a watch." According to Palmer, Lillard had told him that his deafness was brought on when "he was exerting himself in a cramped, stooping position, he felt something give way in his back and immediately became deaf."[39] After examining Lillard's spine, Palmer observed a vertebra racked from its normal position, reasoning that if it was correctly placed, then Lillard's hearing would be restored. It took Palmer thirty minutes to convince Lillard to let him manipulate his spine — thirty minutes to fix seventeen years of deafness.

And thirty minutes to launch the beginning of chiropractic medicine.

Palmer and Lillard's meeting is a creation legend, with several variations on the tale of how the cure and eventual discipline came about. In Palmer's telling, the procedure took place in his office, where he convinced Lillard to "rack" the misaligned vertebra into position "by using the spinous process as a level." In another version, the procedure occurred in an elevator. Lillard remembered it differently. In his interpretation, he had been swapping jokes with a friend in the hallway, when Palmer, overhearing the pair, joined them and slapped Lillard on the back with a book in amusement. Later, Lillard realized his hearing was improved and notified Palmer, who then convinced him to participate in the experiment. Lillard's

FIGURE 4.1. Harvey Lillard, the first patient to receive chiropractic adjustment performed by D. D. Palmer. Courtesy of Special Collections and Archives, Palmer College of Chiropractic.

(alleged) written testimony does not explain what brought on the meeting: "I was deaf 17 years and I expected to always remain so, for I had doctored a great deal without any benefit. . . . I had long made up my mind not to take any more ear treatments, for it did me no good. . . . Dr. Palmer told me that my deafness came from an injury in my spine."[40] He also recalls that he received two treatments from Palmer between January and April 1896 and since then, his "hearing remains good."

There "was nothing 'accidental' about this," Palmer explained, "as it was accomplished with an object in view, and the result expected was obtained. There was nothing 'crude' about this adjustment; it was specific, so that no other Chiropractor has equalled it." Word of Palmer's "deafness cure" traveled quickly as others shared their own anecdotal stories of successful adjustments. The treatment especially resonated with those riding the currents of antiauthoritarian sentiments, who wanted alternatives to what they perceived as "a dangerous, elite medical establishment."[41] People from all over the country thronged to Palmer's office, hoping he could adjust their deafness, or treat an assortment of ailments: influenza, stomach complaints, epilepsy, eye diseases, cancer, and even heart trouble.

Patients expressed their relief and satisfaction with Palmer's new cure. Some wrote to *The Chiropractic* (later renamed *The Chiropractor*), Palmer's journal that documented the principles of chiropractic and his success with patients. Though we cannot ignore the exaggerated or fraudulent nature of some of these letters—after all, many early alternative healers published and edited favorable patient letters to boost their profile, and it was common practice among patent medicine sellers to purchase testimonials from other vendors—these letters do offer insight into how deaf persons experienced their hearing loss and sought treatment. Mrs. E. J. Eis of Pleasant Prairie, Iowa, for instance, had a letter published in 1899, describing how, after being hit hard with *la grippe*, she could no longer hear properly. For seven years, her ears would "break and discharge and erysipelas would set in," leaving her "so miserable that life was a burden."[42] After catching a cold, she ruptured her left ear drum during a coughing fit. She heard about Palmer's treatment, but "had no faith in him," before finally becoming "desperate and was willing to try almost anything." After one treatment from Palmer, her cold and cough disappeared; within two weeks, she "came home cured" and able to hear ordinary conversation where previously her husband had to scream into her ears for her to hear.[43]

Palmer's chiropractic theory borrowed heavily from magnetism, vitalism, spiritualism, and osteopathy, merged with religious overtones to affirm its divinity.[44] Chiropractic philosophy, Palmer proclaimed, is the knowledge of the phenomena of life, a practice that includes moral obligation and a religious duty.[45] He styled himself as a proselytizing medical missionary, turning medical spectacle into evangelism, becoming possessive of his knowledge by maintaining an aura of secrecy. He appeared to have changed his mind in 1897, when a railway accident left him fearful his discovery's immortality. Opening the Palmer School of Magnetic Cure, Palmer offered to teach anyone who could afford the $100 fee; later the school became Dr. Palmer's Chiropractic School & Cure.[46] Students were taught to break fevers, heal infections, cure diseases, and return vision and hearing. Initially, Palmer was secretive about his methods: patients were treated for only thirty seconds and no one was allowed observe his technique.[47] After another accident—this time by automobile—Palmer relaxed his stance on teaching his method. His son Bartlett Joshua (B.J.), a graduate of Wharton School of Finance and Commerce at the University of Pennsylvania, eventually chartered the school in 1907, renaming it the Palmer School of Chiropractic, now the Palmer College of Chiropractic.

As historian Holly Folk argues, Palmer's treatment emerged within a progressive health culture that was driven by an array of social forces, such as industrialization, that converged to create a movement welcome to enterprising persons.[48] The movement blurred boundaries between practices such as chiropractic and magnetism, with its practitioners borrowing, sharing, or stealing ideas and business models while claiming to be originals, as did Joseph Gordon and his wife Nannie, who opened their own magnetic healing practice in Rockford, Illinois, and circulated broadside advertisements nearly identical to Palmer's.[49] In their broadside, *Vital Magnetism*, the Gordons list testimonies of successful deafness treatments, including a boy totally deafened from birth who was "treated in many ways and by many doctors" with no success until meeting the Gordons and placed on the "way to total recovery." Mrs. C. M. Kingsley, the prominent sister of a Rockford grocer, had been deaf in one ear for fifteen years; treatment restored her hearing and her "many friends . . . congratulate her upon her good fortune and the cure [is] accepted as a fine compliment to Dr. Gordon." One of Rockford's "brightest school girls" also successfully had her hearing restored after being deaf since age three, with the "saving of her hearing . . . a great benefit to her finishing her education."

Though Daniel David Palmer received national acclaim for his development of chiropractic medicine, his career came at the cost of intense scrutiny. In September 1902, he was arrested for practicing medicine without a license. He pled not guilty and the charges against him were dismissed a month later. Four years later, he was arrested again for the same crime, but this time, his trial became a test case for legally reprimanding unlicensed practitioners in Iowa. The trial was short, with no testimony by the defense. The prosecution's star witness was Avis Fraser, Palmer's former stenographer; though the defense wanted her testimony struck from the record due to privilege, the judge allowed it. Her testimony, combined with Palmer's writing in *The Chiropractic*, swayed the jury, who returned a guilty verdict—the first conviction in county court on the charge of practicing medicine without a license. Palmer refused to pay the $350 fine and on 2 April 1906 he was ordered to serve 105 days in jail; 23 days into his sentence he was released after his wife paid the fine.

While dealing with his legal troubles, Palmer transferred his assets and the school to B.J.'s wife Mabel, a Palmer graduate and faculty member. Upon his release, Palmer argued with his son about the future direction of chiropractic therapy and their competing attempts to market the field as a legitimate branch of medical science. Feeling Palmer's tainted image would tarnish his promotion of the field, B.J. banned him from any association with the Palmer School of Chiropractic and bought out his father's shares. Palmer in turn angrily accused B.J. of trying to steal the business. The father-son rivalry heightened in August 1913 when D.D. was marching during a Founder's Day parade in Davenport and was struck yet again by an automobile allegedly driven by B.J.[50] Deeply resentful of his son's ascendancy, Palmer died from typhoid fever two months after the accident, at the age of eighty.

Finger Surgery

During the summer of 1923, the *New York Times* and *Time* magazine reported that King Alfonso of Spain had summoned a famous

New York osteopath to treat his fifteen-year-old son, Infante Don Jaime. Deaf and mute following a severe case of mastoiditis (a form of middle-ear infection) and possibly tuberculosis, Don Jaime was judged "incurable" by Spanish specialists who had attempted to restore the young prince's hearing. One surgeon unsuccessfully performed a difficult surgery in 1912 that only worsened the prince's condition. According to the sensationalist newspaper *World*, in 1920 a London-based bonesetter named Dr. May was summoned by the desperate king to perform an innovative surgery. The story, however, proved to be pure fabrication. Another report testified that an unnamed physician recommended the child to be sent to a sanitarium in Switzerland for a full course of the "rest cure." Despite all these failed treatments, Don Jaime was tutored and cared for most of his life by Valencian nurses whose primary mission was devoted to educating the deaf and mute.[51]

Brooklyn-based Dr. Curtis H. Muncie happened to be sailing on the *Majestic*, bound for Europe that summer. Reporters deduced that Muncie was the American osteopath summoned by King Alfonso, likely to perform his famous bloodless and painless "constructive bi-digital intra-aural" technique, otherwise known as the "Muncie Reconstructive Method," or simply "finger surgery."[52] This was a highly specialized procedure of Muncie's own creation that required him to insert his delicate fingers through a patient's larynx to manipulate the Eustachian tube and manually correct aural defects causing deafness. Allegedly, this cure had a 90 percent success rate in otherwise incurable cases.

As reports of Don Jaime's miraculous surgery made headlines, King Alfonso's court sent a cablegram to the Spanish ambassador in Washington, D.C., demanding a correction of the erroneous reports.[53] No physician, Muncie or otherwise, had cured the prince's deafness. Facing accusations of charlatanry and fraud on his return to the United States, Muncie sent out a press release insisting that the stories connecting him to the prince were false and an "unfair and malicious attack" on him and his method, likely planted by enemies of osteopathy who were threatened by his financial success.[54]

He clarified that he had sailed to Europe to treat a prince, but had not named the prince in his original publicity statements, being bound by patient confidentiality. He had never once said—nor ordered anyone to say on his behalf—that he treated the Spanish prince.

Despite criticism, the Don Jaime fiasco put Muncie in the international spotlight. Newspapers and magazines described his finger surgery, proclaiming it as the latest modern cure for deafness. From 1910 to 1960, he was reputed to manage the world's largest otology-osteopathy practice, which occupied the entire twenty-ninth floor of the fashionably exclusive Hotel Delmonico facing Central Park in Manhattan. At his practice, a grand reception room greeted prospective patients: a plush velour couch, high-grade Persian rugs, gilded-framed artwork on papered walls, and a table decorated with fresh flowers and copies of the *American Journal of Osteopathy*. Velvet curtains separated the waiting room from the secretary's office, where an executive desk, cleared of the day's files, remained polished for the next round of patients. Around the corner, the private consulting office invited those seeking Muncie's expert advice, the room furnished only by a desk, a Tiffany lamp, and two sofa chairs. A framed diploma was the only decoration on the wall. There were six other rooms adjoining a narrow hallway: two operating rooms with various equipment and tools, three treatment rooms, and an irrigation and case record room for quick outpatient treatment. Later, one of the rooms would be transformed into a "soundproof" examination room, fitted with an audiometer, tuning forks, an acouphonometer, and other precision instruments, ignoring the fact that with several windows and radiators, the space was hardly soundproof.[55]

Over the course of his fifty-year career, Muncie allegedly treated 1,585,000 cases of deafness using finger surgery. He even had his hands insured for $400,000 and his practice was so successful that rumors swirled he had generated over half a million dollars in profit during the Great Depression.[56] His procedure was an expensive one: the examination alone cost $25, operations began at $200, and postoperative treatment was $10—all mighty sums given that a stay at the Hotel Waldorf in Manhattan cost $5–$10 a night.[57] Yet on any given day, up to two hundred anxious deafened persons lined the hotel corridors waiting for treatment, and Muncie claimed he performed approximately one hundred operations daily.[58]

Born in Brooklyn to Edward and Elizabeth (Libbie) Muncie, Curtis Muncie was exposed to medicine from a young age. In 1896, his parents established the Muncie Sanatorium of Brooklyn, the first homeopathic dispensary; a year later, Edward opened the Surf Sanatorium on Muncie Island, a seaside resort in Babylon, Long Island (the island was depleted in 1933). Libbie was equally accomplished: after obtaining her M.D. in 1891 from the New York Medical College and Hospital for Women, where she specialized in gynecology, she studied orificial surgery at the Chicago Homeopathic Medical College from 1892 to 1895.[59] Afterwards, she continued her surgical studies at Johns Hopkins Hospital before devoting her career to the Muncie Sanatorium, focusing mostly on chronic disease and maternity cases. Some years later, in 1910, Libbie published a book on eugenics and parenthood titled *Four Epochs of Life*.

Curtis Muncie, it seems, was expected to follow in his parents' footsteps, though it appears he initially rejected a career in medicine, deciding instead to study chemical engineering at the Brooklyn Polytechnic Institute. Muncie decided to branch out from his parents' focus on homeopathy to study the principles of osteopathy: the process of physically manipulating the body's bones, tissues, and muscles with limited use of instruments. A health reform movement founded by Andrew Taylor Still in 1874, osteopathy is based on the principle that disease is rooted in a disordered musculoskeletal system that interferes with the nerves and the blood supply. This principle was common among proponents of nineteenth-century medicine who argued that the body was naturally designed to battle harmful threats to health and that aggressive interference would only worsen the illness. Osteopathy would relieve musculoskeletal disorders, leading to improved health without the negative side effects of pharmaceuticals.[60]

Matriculating at the Philadelphia College and Infirmary of Osteopathy, Muncie then transferred to the American School of Osteopathy in Kirksville, Missouri—the first of its kind, established by Still—and graduated with his degree in 1910. He returned to his parents' sanatorium as a general practitioner in osteopathy. A routine examination of an eleven-year-old boy deafened by scarlet fever would inspire Muncie to further specialize in aural surgery. After a standard removal of the boy's adenoids to improve his breathing, Muncie manually examined his nasopharynx with a finger and "found lateral adenoids remaining, which had grown about the orifices of the Eustachian tube."[61] Applying pressure with his finger, Muncie broke the adenoids down and much to his surprise, his finger slipped into the Eustachian tube, thus challenging conventional anatomical findings that the tube was no larger than a goose quill— a discovery that was harshly criticized by other physicians and otologists.

Though the boy's hearing appeared to be "miraculously" restored, the effect only lasted for two weeks. Wanting to develop a more lasting solution, Muncie examined over five hundred Eustachian tubes of patients at the sanatorium from 1910 to 1916, assessing his findings to conclude that the tube *"is larger in life than in death—large enough to be treated digitally."* Having perfected this finger surgery, Muncie treated the boy again, digitally correcting defects in his Eustachian tube and permanently restoring his hearing. His claims of original invention, however, were also challenged: supposedly laryngologist Frank E. Miller first developed the technique as early as 1893 but failed to convince his conservative colleagues of its benefits.[62]

In 1921, Muncie announced two major claims. First, the Eustachian tube is different in size, contour, and tone in deaf persons. Second, Eustachian tube deformity is a universal cause of deafness. He identified eleven types of deafness, ten of which he claimed were curable by his surgery. Only cases of deafness due to nerve degeneration from meningitis were not suitable candidates; nevertheless, he still accepted such patients and claimed he could cure them![63] Furthermore, although Muncie argued that even the slightest tubal deformity could be digitally manipulated to treat deafness, his method could not be taught. Rather, he offered demonstrations for his bi-digital intra-aural technique in more than fifty cities in the United States and Europe, raising his profile as a miraculous healer.

"The patient," Muncie explained in one promotional pamphlet, "comes back from a pleasant sleep with a reconstructed nasal and nasopharyngeal area. No cutting, no surgical shock, but instead, *restoration of the tissue to the normal,* and for this reason, no bad after effects, such as are invariable experienced following instrumental surgical insult. . . . The patient is able to go on to business after the hospital. No hospital care is necessary." With this constructive procedure, he insisted, all original causes of deafness—especially for catarrhal deafness—is removed by "normalizing the nose." This not only corrected the deafness, but also enabled patients to overcome their susceptibility to catching colds and breathe more efficiently, and of course, it established a fully functional and anatomically normal Eustachian tube. Moreover, as Muncie often boasted, the entire procedure was bloodless, painless, and took no more than ten minutes to complete.

Within the osteopathic community, Muncie was celebrated for his technique. He published more than sixty articles in various osteopathic journals and released multiple editions of his self-published pamphlet *Prevention and Cure of Deafness.* Yet his fame was owed not to his professional brethren as much as to his skillful handling of the lay press to promote and popularize his technique. Indeed, his parents had been pioneers of this method, using the press to promote the Muncie Sanatorium in a series of sensational case studies. In June 1914, for example, the *New York Tribune* published an article titled "Knife Cuts Veil From Girl's Mind," describing how Edward and Libbie performed four operations on seventeen-year-old Dorothy Schless, a "wanderer" familiar to many of the city's charitable institutions. Brought to the sanatorium by a probation officer, Schless was examined and diagnosed with a "nervous affliction" due to pressure from a dislocated vertebra. After the operation, the girl recovered her "full mental powers" and was discharged, optimistic that she would find permanent work and become self-sufficient.

As scholars Ravi Swamy and Robert K. Jackler point out, Muncie's familiarity with the power of the lay press likely brought more patients to his practice and spearheaded his procedure as a popular fad. The headlines were certainly tantalizing: "CURE FOR DEAF IN 10 MINUTES! TINY OPERATION." "DEAF MADE TO HEAR BY FINGER SURGERY." "CROWDS WAIT FOR 'FINGER SURGEON' TO HEAL DEAFNESS." "DOCTOR CURES DEAF PATIENTS BY TOUCH." "BLOODLESS OPERATION MAKES DEAF HEAR." And of course, "THE MOST VALUABLE HANDS IN THE WORLD"—for after all, the good doctor's hands were insured.[64]

Testimonials from Muncie's patients proved irresistible to publications with wide circulations. The Associated Press, *National Magazine*, *McClure's Magazine*, and *Sunday Mirror Magazine* all profiled Muncie's practice. Some of the articles, however, came under scrutiny for being transparent advertising. *McClure's*, for instance, was investigated by the AMA after it published a feature profile on Muncie in its May 1925 issue.[65] Written by Viola Roseboro, the article argued that Muncie, a slim and active young man of "seeming exhaustless vitality," was a pioneer in his methods and uniquely successful in his practice. The eleven-page profile covered the range of Muncie's career, from his beginnings at his family's sanatorium to his rise to notoriety after the Don Jaime confusion. Writing to the AMA, William H. Walsh of the American Hospital Association, declared that the article was nothing more than "a paid advertisement, the 'Dr' Muncie is a palpable faker and, that by reason of the collusion of a popular magazine, he is thus enabled to delude many who are handicapped by incurable deafness."[66] Physicians worried that publicity in esteemed magazines such as *McClure's* could legitimatize Muncie's practice, thus misleading deaf people.[67] George E. Shambaugh wrote to the editor of *McClure's* to impugn Muncie as a notorious fraudster whom the public needed to be protected against. The editor, Shambaugh asserted, must have published the article without being aware of the "unscrupulous manner in which those afflicted with incurable deafness were being exploited by this silly finger surgery."[68]

No favorable magazine profile could quell the claims that Muncie

RECONSTRUCTION OF THE EUSTACHIAN TUBES

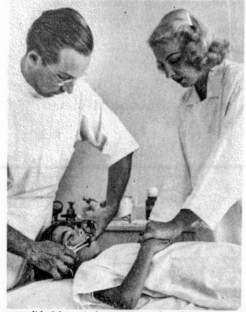

This is accomplished by a single operation without instruments. The sensitive finger of the aurist determines the exact type and extent of the tubal deformity, as well as the state of health of the surrounding tissues, then through a moulding (plastic) process, the deformities of the eustachian tubes are completely corrected. *Here skill cannot be measured by force and time but by dexterity born of accuracy of judgment and complete mastery of technic. No instruments are used.*

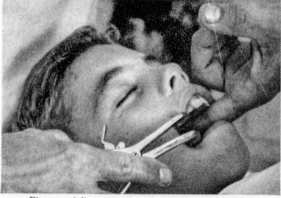

Close-up of Reconstruction of right Eustachian tube.

20

FIGURE 4.2. Demonstration of Curtis Muncie's "finger surgery," with close-up (1930s). Historical Health Fraud Collection, American Medical Association.

LOS ANGELES RECORD

Attacks Deafness Wi h Fingers—

FIGURE 4.3. An article promoting Muncie's surgery. Note the number of patients lined up in the corridors of Hotel Waldorf waiting their turn (c. 1930s). Historical Health Fraud Collection, American Medical Association.

was a palpable faker who deluded his patients with his unscrupulous fad procedure. The AMA insisted Muncie's technique lacked any sound scientific support and declared it as the worst deafness cure fad of the 1920s and 1930s. Indeed, in 1942, the association's journal editorialized that "Muncie may well rank among the leading charlatans of our time," as his aggressive advertising, promotions, and boastful pretensions were all characteristic indicators of faddish medical quackery. Such exposés, however, only bolstered Muncie's reputation.

Deaf persons from across the country wrote to the AMA inquiring about the benefits of Muncie's treatment, asking for more information, and sharing their own experiences. Most of the letters expressed dissatisfaction and regret over lost funds from $3,000 to $2,000, but more so, over the disappointment of the treatment's failure to restore hearing. Some patients continued treatments for a year or more, even without any evidence of improvement. Winifred H. Blanchard, for instance, was treated for over a year. Muncie's promises that she would never need to lipread strengthened her trust in his method, even as her condition worsened.[69] "I am somehow doubtful," wrote Mrs. H. Hallenstein, "about the absolute honesty and sincerity of this doctor and would like your opinion."[70]

Some of the letters were published in *Hygeia*, the AMA's lay magazine. Established in 1923, the magazine was designed to educate the public about health matters in nontechnical language; mothers and teachers were the primary readership target. In one reply to a reader, *Hygeia* remarked that Muncie's claim that catarrhal deafness could be cured with his procedure was "too ridiculous to even discuss. He is considered unscientific and his treatment has no virtue."[71] In another issue, W. H. B. of Chicago received a third-place prize of $5 for his entry in *Hygeia*'s contest on the "Worst Health Fad or Fraud I Ever Fell For." Swept by the promise of permanent improvement, the author had borrowed funds from a relative to try out Muncie's finger surgery, forking over a total of $1,500 for the procedure and follow-up treatments. "How could I have been so imposed upon?"

he continued. "I was desperately anxious to save my hearing, his office was imposing and filled with a fine appearing class of men and women, and I saw copies of [medical journals] . . . and supposed he was at least honest and sincere."[72] Despite the author's insistence that he obtained no benefit from the procedure, Muncie asserted that everything was in order and that the condition was actually improving—even though evidence demonstrated the author's hearing was getting progressively worse. Physicians also wrote to the AMA narrating patient experiences of being swindled by Muncie's "fraud practice." In one instance, a father took his two-year-old deaf-mute son for Muncie's treatment from October 1923 to June 1925, making between fifty-five and sixty trips from Worchester, Massachusetts, to New York City, and spending upwards of $1,000.[73] At the end, the father realized that no benefits could be obtained for his son.

Not all readers agreed with the AMA's critical stance of Muncie. One of Muncie's discharged patients, deaf for fifteen years and no longer under treatment, urged the AMA to include various perspectives from patients regarding this unconventional treatment: "I, like any other person living, did not care under what society he practiced if he could help me."[74] Another patient wrote to the AMA denouncing their attacks on Muncie. Stories of successful finger surgery, wrote Mary O'Reilly, are "for the people whom he has made happy and restored confidence in themselves to tell the world what a genius he is." It is outrageous, she continued, that Muncie "is accepted in Paris, London and Glasgow for his true worth, and condemned by his own country, America, the Land of Friendly Relations." The AMA, O'Reilly declared, should stop its baseless attacks on osteopathy and attempts to weaken confidence of those believing in alternative treatments. "Cure the people as you claim you can do, and they will not seek a Doctor of Osteopathy. We are a nation of thinking people, and refuse to respond to the crack of a whip, to be herded like cattle." Mrs. L. M. Hanks also advised the *Volta Review* (a national magazine for the hearing-impaired) to be careful in classifying practitioners like Muncie a "pseudo-physician." Hanks was deaf for twenty-five years and had tried numerous treatments, including a visit to esteemed otologist Harold Hays in New York City and tried a Port-O-Phone hearing aid, but nothing restored her hearing until she visited Muncie's office. At the office, she witnessed "old cases of twenty years standing, but without middle ear degeneration, regain normal hearing, and have seen those whose deafness was of short duration, cured."[75]

Such letters reflect how deaf persons contextualized their treatment choices despite the medical profession's disdain for procedures that did not conform to perceived scientific and professional standards. But for patients, the bottom line was that the treatment needed to work—and some claimed it *did* work. Despite the AMA's criticism, Muncie's career flourished. Swamy and Jackler argue that it was Muncie's "uncanny ability to maximize his exposure in the public limelight that drowned the accusations of the medical establishment over his lack of sound scientific proof and unethical behaviour." It was not criticism that did his career in, however, but rather his failure to pay his taxes. In 1939, Muncie was arrested and indicted on five counts of income tax evasion totaling $159,280 and released on a $1,000 bail. Three years later, he pled guilty, and was sent to jail and ordered to pay a fine of $100,000.[76]

After his release from prison, Muncie discreetly continued his practice. Like D. D. Palmer, Muncie expanded his business by apprenticing his son Douglas, who had completed his degree in osteopathy in 1942. Trained in his father's secret maneuver, Douglas eventually established the Muncie Institute for Hearing in Miami, and later a satellite office in Las Vegas, providing a "non-surgical" operation to cure congenital deafness in children. Audiometers showed that the hearing of some patients immediately increased about 50 percent, increasing up to 71 percent over several months following the treatment, proving that finger surgery had "worked wonders."[77] Like his father, Douglas Muncie was perceived as a problem for otologists, denounced as "an out and out fraud and quack."[78]

The Window Operation

A forty-four-year-old patient, stone-deaf for half her life, lay on the operating table. Two dental drills covered with sterile linen sleeves hang over the table, prepared and ready. The surgeon, a wispy, small man with a bushy pompadour adding inches in height rather than presence, signaled the anesthetist that the procedure was about to begin. Turning on his headlamp, he focused a powerful, pencil-thin stream of illumination directly on the patient's right ear. Slowly, but steadily, he began the first incision on her outer ear, creating an opening the size of a small coin. With one of the drills, he penetrated through the mastoid bone, proceeding for half an hour until the bony labyrinth of the inner ear was exposed. This would prove to be the new "window" that would restore sound vibrations to enable his patient to hear.[79]

First announced to the otological world on May 1938, Dr. Julius Lempert's three-hour fenestration operation was a bold innovation for treating otosclerosis, a hereditary condition that causes progressive hearing loss. Otosclerosis occurs when sponge-like bone tissue grows around the ossicles—the three bones of the middle ear—due to an overproduction of calcium, effectively immobilizing them and preventing the proper passage of sound vibrations necessary for hearing. Prior to the 1930s, the condition was deemed incurable by leading otologists, who agreed that only a hearing aid could assist in amplifying hearing in a patient. Yet this did not stop numerous offers of cures, many of which came from quack practitioners who promoted their procedures through advertisements and mail-order flyers. As surgeon Howard P. House, founder of the House Ear Institute in Los Angles observed, "Perhaps no other group of individuals has been subjected to more treatment without benefit than has the unfortunate patient with otosclerosis. Repeated Eustachian tube inflations, nose treatments, nose and throat surgery, ear drum massage, various diets, endocrines, vitamins and minerals have all been put through their paces without noticeable benefit."[80]

British aural surgeon Joseph Toynbee first made the association between hearing loss and otosclerosis in 1841. Through a series of dissections on the temporal bones of the ear, he discovered that when the stapes was immobile, deafness was usually the result. Otologists thus began a search for maneuvers to remove the obstruction to sound vibrations—including surgically creating a new opening into the labyrinth to bypass the bone. Early operations yielded a substantial restoration of hearing immediately after surgery but produced disappointing, if not disastrous long-term results.[81] After a few weeks or months, the bony growth often returned or else patients died from secondary infections.

The threat to patient health from this "possible" cure was so damaging that, in 1894 at the International Congress of Medicine in Rome, renowned otologists Adam Politzer, E. J. Moure, and Vincenzo Cozzolino condemned all surgical attempts to treat otosclerosis. Otologists reprised this condemnation at the next congress in 1900, when they agreed to reject all forms of invasive surgery for deafness. For the time being, the deafened were protected against dangerous, ill-fated surgeries for their hearing loss. At the same time, the period saw a tremendous rise in patent medicines and innovative devices marketed to otosclerosis patients.

The condemnation did not mean attempts to treat otosclerosis ceased altogether. In fact, quite the contrary. In 1916, Stockholm otologist Gunnar Holmgren and his assistant Carl Olof Siggesson Nylén conducted careful observational trials of safe surgical techniques to treat of deafness caused by otosclerosis. Aided by Nylén's creation of a binocular microscope, which allowed for proper illumination to make microscopic surgery possible, the team was able to observe the possibilities of treating the condition by fixing the stapes footplate directly.[82] With the microscope, Holmgren discovered that even if a new fenestra—that is, a window—was created, the bony growth eventually returned and closed it. It was impossible to predict when this would occur, or in which specific cases of otosclerotic deafness.[83]

Armed with this study, otologists attempted to devise new maneu-

FIGURE 4.4. Julius Lempert. *The New York Physician and American Medicine* 56.6 (July 1961). Courtesy of the New York Academy of Medicine.

vers to prevent this phenomenon. In 1924, French surgeon Maurice Sourdille visited Holmgren and was impressed by his work. Five years later, he modified Holmgren's technique, creating a three-stage operation for creating and sealing the new fenestra. By 1935, he had performed over 150 of these "tympanolabyrinthopexy" operations; he demonstrated his technique at the Royal Society of Medicine and New York Academy of Medicine in 1937. These advancements, and the increasing technological improvement of vacuum-tube hearing aids, led Norton Canfield, a Yale-based otolaryngologist, to declare that a "new phase of otology has been ushered in."[84] And it was at the academy that Lempert learned about Sourdille's technique and became inspired.

Born in Lublin, Russia (now Poland), Lempert grew up in a time of rising unrest. After joining an antigovernment street demonstration, he was pursued by mounted Cossacks and was captured hiding under the bed in a neighbor's house.[85] Fearing for his safety, in 1905 his family immigrated to the United States and settled on the Lower East Side of Manhattan; fourteen years later, they would receive their American citizenship. Desiring to be a physician, Lempert enrolled directly from high school to the Long Island Medical School. After graduation, he toured the wards of the Manhattan Eye and Ear Infirmary, and the New York Infirmary, even though he lacked a formal residency. To sharpen his surgical skills, Lempert took many overnight emergency cases at the infirmary, but by 1924, he had overstayed his allotted visiting time and was asked not to return.[86] With no real prospects of a position in a New York institution, Lempert opened his own practice, the Lempert Institute of Enadural Surgery, in a six-story red brick building with sandstone griffins squatting in the front, next to the Episcopal Church of the Resurrection and near Lennox Hill Hospital. To raise his professional profile and increase his patient roster, he offered to kick half of any fee collected back to physicians who referred patients to him. This practice certainly made him one of the busiest otologists in the city, but it also ostracized him from his brethren, who frowned on the degrading practice.

A year after learning Sourdille's procedure, Lempert introduced his own version that reduced the three-stage operation to one, which removed the incus to create a large window and secure it against closure. He named the procedure the "Lempert Fenestra Nova-Ovalis Technic," or simply the fenestration operation, and announced it at a meeting of the American Otological Society. Almost immediately, the procedure catapulted Lempert from obscurity to world fame: patients from all over the world arrived at his institute seeking care, and eager otologists—both young students and seasoned surgeons—visited him to learn more. In the midst of World War II, Lempert trained nine or ten doctors from the United States, Europe, Australia, New Zealand, and South Africa every six weeks, and performed nine fenestration operations on any given day, with each operation taking two hours. By the end of the war, Lempert had performed upwards of three thousand fenestrations.

The eventual publication of Lempert's article describing the operation stimulated interest among American otologists, as well as controversy about the dangers of it being another deafness fad.[87] Otosclerosis was not a condition that was diagnosed lightly, for it was generally recognized to be incurable, much to the disappointment of patients. Skeptical otologists even claimed that Lempert offered nothing more than another worthless procedure with no real expectations for improving the success rate of hearing restoration in patients. Otolaryngologist Emanuel Mann Josephson, for instance, wrote one of the most derisive reviews of Lempert's operation. Describing the operation as "mayhem and human experimentation," Josephson argued that the supposed "cure" was a vicious fraud deliberately perpetuated by organized medicine upon "the pathetic victims of deafness" and promoted to bolster the incomes of "boss otologic specialists that had been cut severely by depression and by technological improvements." Originally circulated as an anonymous pamphlet, the remarks were eventually printed as an appendix in the second edition of Josephson's iconoclastic book, *Your Life Is Their Toy: Merchants in Medicine* (1948), a conspiracy-laden work on the medical elites' agenda to retain a monopoly on health care.[88] "This fraud," wrote Josephson, "was the chief product of the multimillion fund for research on otosclerosis raised by the bosses banded together in the American Otological Society." No procedure, past or present, he continued, has been exploited more ruthlessly than the Lempert fenestration operation, and Lempert himself "found in these experiments on the deaf a royal road to notoriety, a means of rehabilitating his fortune and of ingratiating himself with the speciality bosses."[89] The AMA responded that while Josephson seemed convincing, they did "not consider him to be enough of an authority" for his arguments to be taken seriously.[90]

As was characteristic of popularized new deafness cures, despite criticism from the medical profession, the media assiduously reported on Lempert's procedure, remarking on his "new high in surgical artistry" that improved over time: when first introduced, Lempert could only restore hearing in 40 percent of cases, but after adjustment the procedure succeeded in 60–70 percent of cases by 1941, and in 80–90 percent of cases by 1952.[91] "The miracle of making the deaf hear has been accomplished, apparently permanently, in 126 out of 150 cases," noted the *New York World*.[92] Widely referred as the "window operation," the procedure was discussed in feature articles in Sunday newspapers and by popular columnists like Damon Runyon, Logan Clendening, and Ed Sullivan, who also defended Lempert's credibility.[93] Local papers and popular national magazines like *Saturday Evening Post, Time, Parade, Reader's Digest,* and *Hygeia* printed headlines declaring "NEW HOPE FOR THE DEAF," though their assessments were somewhat exaggerated. *Hygeia*, for instance, published an article with that headline, which was later condensed, reprinted, and given wider circulation in *Reader's Digest*. The article claimed that Lempert's technique improved hearing in 98 percent of cases — giving the impression to deafened persons that the operation was a panacea for *all* forms of deafness.

Lempert's backing of several Broadway shows invited further media coverage, as did his flourishing in New York high society (where he was introduced by his wife, Florence Kennedy of the Ziegfeld Follies) — hosting drinks at the famed Lindy's or the 21 Club, where he was known to all, unwinding with Havana cigars, surrounded by surgeons and fellow society elites. Feature stories on Lempert's more high-profile patients — including Leonard Firestone of the tire company and actress Margaret Sullavan — continued to throw a celebrity spotlight on the window operation.[94]

Concerned about the media's tendency to raise false hopes in deaf citizens, leaders of deaf and hard of hearing organizations and schools worked to clarify the merits of the window operation. Deaf educators advised interested persons to seek a respectable ear specialist and to be wary of quack institutions that masqueraded as reputable. The January 1946 issue of *The Pelican*, the magazine of the Louisiana State School for the Deaf, for instance, did not claim any authority to advise parents of deaf children on the value of the operation. To anxious parents desiring advice on curing their child's deafness, the school urged: "Go slowly. Go to a reputable ear specialist." They advised caution, for even a prominent ear specialist, "himself very hard of hearing because of otosclerosis . . . preferred not to take a chance on the operation on his own ears."

The growing number of stories recounting successful hearing restoration further solidified the need to clarify which patients could benefit from the treatment. Indeed, Lempert himself explained in interviews that the operation was only for deafness caused by otosclerosis, and only if the auditory nerve was in good condition. He compared the operation to "fixing a telephone receiver in the presence of an actively working telephone wire": if the line is dead (i.e., the auditory nerve), then no sound could be obtained, no matter how successful the repair on the receiver.[95]

The headquarters of the American Hearing Society (AHS) in Washington, D.C., also received a flood of inquiries regarding Lempert's operation, as did the New York League for the Hard of Hearing.[96] Since the AHS did not give out medical advice except through its national newsletter *Hearing News* and other published leaflets, it decided to publish a special feature article with prominent otologists to answer some of the most common inquiries.[97] Aiming to provide information about the stages of the procedure, the AHS wanted the otologists to cut through muddled anecdotes and sensationalized news stories to clearly advise potential patients. They published answers without medical jargon, and a key segment focused on the operation's ability to restore hearing. Otologist Gordon D. Hoople clarified that the operation could not give perfect hearing; at best, only a practical hearing aid would allow patients to hear religious services and lectures and to participate in group conversation. Not every patient, however, would experience this maximum result. Other otologists replied that the operation would provide up to twenty to thirty decibel gain, or hearing restoration as high as 70–75 percent. All noted that the procedure was *only* for patients with otosclerosis, and that the poorer ear was chosen for surgery—this left the better ear untouched and available for the use of a hearing aid if necessary. The second ear would be operated upon only if the treatment of first ear was successful for at least two years.

The operation did not benefit all patients. Fifty-five-year-old Mrs. Fay Sobel, who suffered with ear noises and partial deafness for fifteen years, wrote to Hallowell Davis and S. Richard Silverman at the Central Institute for the Deaf in St. Louis, begging for assistance. "When this Fenestration Operation came into being," she wrote after spending years searching for relief, "it was a dream come true. At last something had been discovered to stop ear noise."[98] She visited a specialist who assured her that the operation would help; unfortunately, it failed and her ailment worsened, even despite three follow-up procedures. "The racket is driving me Franatic [*sic*] I can't eat, sleep or even think straight," Sobel exclaimed. "It is like dogs barking, bells ringing, and drums pounding continuously I am afraid that I shall lose my mind if these noises do not subside soon." Pleading to the doctors, she offered them $5,000 for any operation that could help. "That is how desperate I am." Nor did the operation benefit Arthur Wolff of Chicago, whose severe tinnitus brought him to Lempert's office in New York. Disappointed by Lempert's explanation that surgery would not improve his hearing impairment and only slightly improve his tinnitus, Wolff wrote to *Hygeia* that he was unable to afford the expense of receiving treatment in New York and inquired whether there were other specialists in Chicago. The surgery, he emphasized, "was worth a few thousand dollars" if it could cure the tinnitus. He also noted that "I have seen more than a dozen patients at Dr. Lempert's office, where I have been twice for examination. All these patients had the window-operation and all say that it is just wonderful; nobody would ever believe that they have been hard of hearing."[99]

Malpractice suits also threatened Lempert's operation and the integrity of the fenestration operation. One suit claimed that two years after the procedure, the patient lost all residual hearing and went completely deaf; Lempert paid $12,000 in settlement. Another claimant, Mrs. Racie B. Sherry of Chicago, was awarded $24,000 by a jury in connection with her $125,000 malpractice suit against Lempert. Sherry charged that Lempert failed to follow orthodox guidelines as outlined by otologists, which resulted in permanent impairment of her hearing.[100] Despite the media's tendency to favorably report on the operation, few reported the malpractice suits against Lempert—or if they did, they presented them with no editorial comment.

Nevertheless, by 1950, over 15,000 fenestration operations had been performed in the United States. The procedure was far from bloodless, painless, or without danger. Only patients diagnosed with otosclerosis were recommended for the operation, but the diagnosis itself was extremely difficult. Otologists trained under Lempert sowed further confusion by offering their own criteria for selecting patients. The rate of misdiagnosis was high, as patients frequently had other associated symptoms, such as trauma, childhood disease, or upper respiratory infection, which was more likely to cause hearing loss than otosclerosis. Furthermore, while the procedure promised great advances, it also risked nerve degeneration after a certain age, risking permanent deafness in patients who were only partially deaf or had residual hearing. It was especially perilous to operate on children, so otologists advised that for best results, patients should be between fifteen and forty years old—preferably under twenty-five—showing no signs or history of middle-ear disease or nerve deafness. By following these criteria, the success rate of hearing restoration in patients could be 80 percent.[101]

Even if fenestration surgery only benefited between 1.5 percent and 5 percent of the hard of hearing population in the United States, curing deafness through surgical intervention was radical and innovative.[102] It held tremendous benefits for individuals with hearing loss who had gone through the entire roster of nostrums, emetics, oils, artificial eardrums, vibrating massages, and light ray therapy, all of which failed to restore hearing, even partially. Yet, as otologists insisted, similar benefits could be obtained by hearing aids without risking permanent damage to the auditory nerve. One 1948 survey, however, found that for some patients, the discreetness of surgery and erasure of obvious signs of hearing impairment were preferable to wearing a hearing aid. While the success of fenestration operations, the survey reported, would "reduce the previously exaggerated hearing aid market in the U.S., the psychology of the deafened is such that most hard of hearing individuals, if a choice is available to them, will prefer a fenestration operation to a hearing aid."[103] A twenty-six-year-old woman, for instance, whose hearing had been declining since age fourteen and had only 40 percent usable hearing, wrote to prominent surgeons about her dilemma of choosing an operation or a hearing aid.[104] And as Logan Clendening said in his column *The Doctor's Corner*, those who realize that they are gradually becoming deaf "think they are going to get better. They can't believe it is going to happen to them. There must be a cure somewhere, in this modern age of miracles."[105]

The media attention not only popularized the surgery among the deafened public, but also elevated its faddism, encouraging more otologists to consider offering it to their otosclerosis patients. In addition to the students trained under Lempert, established otologists inquired for details about the operation and discussed its technical aspects in order to incorporate it into their own practices. At Washington University School of Medicine, Theodore E. Walsh often received referrals for fenestration patients, including one in 1952 who had a history of five or six years of progressive loss of hearing and was diagnosed with clinical otosclerosis. Since the loss on her left ear was worse than the right (52 db loss at threshold, compared to 42 db), Walsh decided to operate on the left ear only, and it appears the procedure was successful.[106] Otologists were especially concerned with the question of whether the technique was able to completely open the bony canal and *keep* it open for improved acoustic gain. Indeed, in 1941, otologists specializing in otosclerosis surgery decided to tabulate their results to track successful trends: 616 cases were submitted by twenty-two surgeons. Of those cases, 300 were done by Lempert, with George E. Shambaugh Jr. trailing with 103 cases.[107] The survey concluded that most cases of fenestration surgery was successful for treating otosclerosis.

The fenestration operation did not remain static. Much to Lempert's disdain, otologists departed from his precise surgical technique to make their own modifications, beginning with Shambaugh Jr., his first student. Shambaugh Jr. recounted the first moment he watched the procedure and as the horizontal canal was opened, the patient cried out, "I can hear! My God, I can hear!" "Chills ran up and down my spine," Shambaugh Jr. recalled. "It was the single most exciting and impressive moment in my entire medical career. . . . The moment was miraculous and never to be forgotten."[108] The closing of the fenestra, however, remained an issue for Lempert, as it did for Sourdille. Experimenting with rhesus monkeys, Shambaugh Jr. discovered that after the operation, bone dust particles remained and were potent enough that new bone would form to close the fenestra. He then introduced suction irrigation to ossify the fenestra and remove all bone dust particles; this greatly reduced the likelihood of reclosure.[109] After the press announced the modified technique, Shambaugh Jr.'s office was flooded with new patients. Lempert was not happy. "Dr. Lempert, who had been most cordial and helpful until then," Shambaugh Jr. recounted, "became exceedingly hostile toward me for deviating from his technique. He felt that I was deprecating *his* operation, which he jealously valued as a mother would her child."

Otologists successfully performed the procedure in the U.S. as well as in Denmark and Norway, but by the early 1950s, it ceased to be the primary surgical avenue for otosclerosis after Lempert's rival Samuel Rosen developed a less invasive procedure to remobilize the stapes through the ear canal.[110] Further improvements to stapes remobilization were performed by John Shea Jr., who replaced the stapes with a prosthesis (stapedotomy). Fearful of irrelevance and fiercely overprotective of his creation, Lempert refused to acknowledge either operation; thus, his practice suffered. As British otologist Terence Crawtowne explains: "Lempert finally achieved the pinnacle of professional success and recognition only to have the limelight snatched away as stapedotomy superseded fenestration surgery. Perhaps this hastened the cerebral deterioration leading to his invalidism and eventual demise." Forced to sell his hospital and facing problems with the Internal Revenue Service, Lempert fell into a deep depression in 1961. Diagnosed with early-onset dementia, he moved to a nursing home outside of New York, where he died on 14 December 1968. Nearly all his students—including George E. Shambaugh Jr., Howard P. House, Gordon D. Hoople, and Philip Meltzer, who had become respected surgeons in their own right—attended his memorial service. A tenacious, colorful character, "nobody who met Julius Lempert could ever forget him," Terence Cawthorne fondly recalled.[111]

Proportions of Faddism

Fads and fancies, historian John E. Vacha tells us, are two sides of the same coin. Fads are temporary fashions enthusiastically followed by a group, whereas fancies are aspects of imagination or inclination: "Fads, then, are individual fancies popularized, though few fancies reach the mass proportions of faddism . . . [and] two prerequisites for the making of fads are leisure, in which to pursue them, and mass media, by which they are spread."[112] New medical fads emerge all the time as well, mainly because so many diseases are incurable or chronic and can only be palliated. These fads even offer genuine surgical treatment, but overreliance or misapplication often raise skepticism in their value. Yet not all fads disappear; some evolve and are reborn in new guises to take their rightful place within faddism: variations of fenestration surgery and osteopathic and chiropractic treatment for deafness are still practiced today.[113]

There were many avenues for deaf people to learn about the latest deafness cure and push it towards faddish heights. In addition to popular newspapers and magazines, "little paper family" publications—periodicals written and printed by pupils in residential deaf schools—were one essential site for information. Designed to entertain and enlighten, these papers kept subscribed readers informed about student, school, and community life, while creating a national network of information exchange. They became places to ask about and debate deafness fads, if only to inform or reassure parents of deaf children that a possible cure was available. *The Silent Worker*, which was printed at the New Jersey School for the Deaf, for instance, announced in the January 1913 issue that although several students had been taken away to be "cured" in the latest (unnamed) fad operation, "all have returned, a circumstance that makes us not altogether hopeful. . . . Of all the ills that flesh is heir to deafness appears to be one of the hardest to reach."[114] Eager patients also wrote directly to medical experts, outlining their case histories and inquiring whether the latest procedure could benefit them. One forty-six-year-old woman with otosclerosis, for instance, experienced worsening high-tone hearing loss after the birth of her two children; she had two fenestration operations and inquired whether a plastic tube prosthetic could lessen her dependence on a hearing aid.[115]

Elsewhere, social advocacy groups such as the New York League for the Hard of Hearing and its national umbrella organization, the American Society for the Hard of Hearing, printed newsletters that included readers' stories of newfound fad cures. They also corrected misconceptions based on media reports on fads and invited affiliated otologists to clarify the pros and cons of surgical fads. Indeed, the society's headquarters was frequently overwhelmed with inquiries every time a mainstream news story reported on a "breakthrough" operation to cure deafness. These newsletters provided deafened persons opportunities to learn about the latest deafness treatment while also allowing themselves to convey their fears and worries about being disappointed over treatment failure. Writing directly to otologists, editors, or reporters was another approach for keeping informed.

The publicity surrounding the creators of new deafness procedures also spearheaded a fad obsession. Julius Lempert, Curtis H. Muncie, and D. D. Palmer all romanticized the thrilling nature of their cure, in which a patient laid upon a table and drifted off to sleep, only to awaken miraculously with their hearing intact. By emphasizing the isolating facets of deafness, these men made direct appeals to promises of cure: they asserted their glamorous media presence by providing dramatized stories of patients saved by their skillful hands in a safe and easy procedure. They kept their names in wagging tongues, ensuring that as long as they received attention, so too did their deafness cure, thereby increasing its spread and reaching more potential patients.

5

Edge of Silence

Now, Vanity is working for us. Now hearing aids . . . are part of the personal appearance market, sparked by the biggest single force in the American market. Pride, prestige, vanity—the desire to be at one's best, to look one's best—the desire to be modern, to be smart, to be in the swim, to look important, handsome, dignified—the Distinguished Look—with the world's most attractive hearing aid—a new, powerful key to open up the *entire* hearing aid market.

Zenith film, "The World's Most Attractive Hearing Aid"[1]

I find this Madison Avenue approach to marketing hearing aids offensive, if only by implication. Hearing aids are not merchandise. They are, to a great many millions of people, an absolute medical and social necessity— with the unfortunate status right now, of a considerable luxury. This should not be so.

Testimony of Nanette Fabray, Congressional Hearing
on Hearing Aids and the Elderly, 1968[2]

The film follows the lives of two people—a middle-aged businessman and a teenage girl—struggling to find solace in an increasingly noisy world. They furrow their brows as sounds get muffled and tilt their heads to catch the meanings of sentences that taper off at the end. As the narrator tells us, the protagonists hoard themselves behind an internal psychological barrier made of failure and self-pity. It is this barrier that dooms their chances at success and happiness, creating a downward spiral toward depression. We follow the girl to the dramatic climax at a bridge early one morning. She had enough of life and given up on her dreams, until the man passes by and stops her from jumping into the river. "You're just sorry for me," she says, as he insists there were better solutions for relieving her misery; after all, he too, was suffering in silence. We then are told the two found guidance from a physician and inspiration from "a career woman" and are placed back on the path of successful living. They have come to realize the solution to happiness was a simple thing: a hearing aid.

The Edge of Silence, a sixteen-millimeter film produced in 1955 by Telefilm Productions and sponsored by Zenith Radio Corporation, describes a purportedly common perception held by hearing-impaired persons: that the psychological barrier separating them from the world of sound is oftentimes so overwhelming that it is better to die than to live in deafness. "Perhaps this contagion of sympathy," wrote film reviewer Cecile Starr, was the deciding factor for judges gathered in Chicago for the Film Council of America's Golden Reel Film Festival.[3] *The Edge of Silence* beat out three thousand documentary, educational, and sound slide-films to receive a Golden Reel award in the category of "Health & Hygiene."[4] Ideal Pictures Inc. distributed the thirty-two-minute film to high schools across the country. They also offered copies to physicians who were concerned about their patients, to social or civic organizations, or to anyone writing in to request a screening copy. As Zenith's marketing team asserted, the film should be shown to everyone, everywhere. Of vital interest, of course, were the nation's approximately fifteen million hard of hearing persons.

The film was also an ideal opportunity for Zenith hearing aid dealers seeking to increase sales: "*Never* has there been an opportunity . . . to put across the story of better hearing with such *impact*, and yet do it in so *entertaining* a way. Here's a realistic, completely enjoyable film that . . . *shows* . . . *tells* . . . how the magic of electronics has made it possible for the troubled, hard-of-hearing people to regain their confidence, and lead normal, successful lives."[5] As dealers knew, to sell a hearing aid to a skeptical and resistant customer base required appealing to their sense of isolation and vanity. Products were only one step. Selling a lifestyle, one that unburdened family and colleagues, was a far more powerful strategy. Zenith did not simply sell hearing aids—they sold "Living Sound."

Hearing aid technology underwent a dramatic shift in the 1930s, when subminiature vacuum tubes made it possible to combine the amplifier and microphone into a single pocket unit connected to wire ear receivers. These new models shifted from being portable devices to *wearable* ones: a user could now wear hearing aids on their clothes or keep them in pockets, rather than carrying them in accessory bags, though most models still required separate battery packs. Known as multi-paks, these hearing aids were technologically superior to carbon electric aids, but users often found them frustrating: the struggle to conceal wires, coupled with poor acoustic quality and clothing noise interference, worked against the device's benefits. Hearing aids, users insisted, *revealed* their deafness rather than concealed it. No matter how much manufacturers boasted about the latest technological advancement—including higher gain and superior microphone—they could not create a device that would allow a user to pass as hearing.

Indeed, as one hearing aid user said, "The majority of deafened people probably do not know what an automatic volume control is, nor its value in protecting the ear; in other words, it is not a good selling point."[6] While automatic volume control is necessary for dealing with the "material world where doors slam, hammers bang, auto horns blare," technical improvements could not address a hearing-impaired person's feeling of isolation or unwillingness to wear conspicuous aids. This of course, made it difficult to increase sales. Thus, during the late 1930s, hearing aid companies introduced new advertisements that shifted focus from technical advancements to individual responsibility: *not* wearing a hearing aid was *more* conspicuous, for no matter how well a user concealed their aid with crafty fashioning, the strained "deaf face" gave away their deafness. The deaf face was so discernible that magazines for deaf and hard of hearing people frequently featured articles advising readers how to manage it. The February 1939 issue of *Better Living* magazine (which

FIGURE 5.1. Acousticon advertisement making the connection between poor hearing and health problems (1949). Hallowell Davis Collection, Bernard Becker Library, Washington University School of Medicine.

was sponsored by Sonotone), for instance, editorialized the importance of "Speaking of Appearances." "Let's be honest with ourselves about this matter of conspicuousness," the author writes. "It's a true picture of the difference when you wear a hearing aid—being able to sit back, relaxed and natural, being able to hear what's going on without straining a ligament in your otherwise handsome countenance."[7]

By the 1940s, hearing aid firms employed new advertising strategies to appeal to resistant consumers. Irving Schachtel, president of the Sonotone Corporation, faced this problem by first addressing the prejudicial factors facing the deaf, and the emotional reactions of shame and inadequacy that obligated them to conceal their deafness. One especially troublesome issue was that advertisements for hearing aids used language similar to that of patent medicines—phrases such as "invisible," "hidden," and "concealed," which undermined the technological benefits of hearing aids by inadvertently placing them in the same category as (alleged or known) quack deafness cures. Removing these terms and replacing them with references to "individualized fitting" that provided precise audiological diagnoses was a preferable strategy for personalizing a user's "good hearing."

New marketing strategies thus needed to emphasize that wearing hearing aids was a social responsibility: "THERE IS STIGMA TO BEING DEAFENED—AND NOT DOING SOMETHING ABOUT IT."[8] Such messages became more compelling when manufacturers combined two innovations emerging out of World War II—printed circuit boards and button batteries—to produce even smaller compact models. With these advancements, the batteries, amplifier, and microphone could all be combined into a single unit, the monopak. Their smallness then became the chief selling point of hearing aids, especially as units could be built into headbands hidden in the hair, pendants, tie clips, watches, fountain pens, or even jewelry.[9] Advertisements juxtaposed these aids next to a pack of cigarettes, a deck of playing cards, or a cosmetics mirror. "At last," the Aladdin Company declared, "HEARING HAPPINESS" could be gained from their new and improved vacuum tube hearing aid. This aid was "Inconspicuous but Efficient," with less than six inches of cord showing, and the microphone discreetly tucked away behind the tie or on the lapel. This compact "Unipak Amplifier" can be tucked away in "breast or side pocket as handily as a cigarette case."[10] Maico unveiled its "Hear-rings," jeweled covers for receivers in 1948, as did Sonotone with its "Sono-Charms." Even the cosmetics company Max Factor created a wig for disguising hearing aids.

While advertisements portrayed aspirational social realities, it was hearing aid dealers who propagated these messages directly to consumers. Nearly every hearing aid was sold and fitted by dealers. In the early days of the industry, they were responsible for signing up potential customers—referred to as "prospects"—and selling products through marketing schemes like "free trials" and "guaranteed placements." They sent out personalized form letters and pamphlets to prospects, cold-called customers, and placed advertisements in local newspapers offering their services. Most dealers worked outside of the direct supervision of management and relied on commission sales. By the 1940s, the estimated number of dealers in the United States grew. Some entered the business by establishing their own firms, as engineers Eugene F. Peterson, Orrin W. Livingston, and John H. Hutchings of Schenectady, New York, did in 1940 with their National Electronics Inc., which sold and distributed Aladdin hearing aids.[11] An engineering background was not a prerequisite for dealers; most learned on the job or through the mandatory correspondence course from the company whose products they sold.[12]

Dealers were more than primary sellers. They were an integral aspect of a network of engineers, manufacturers, distributors, otologists, and advertisers that were collectively referred to as "Deafness Fighters," primarily men, situated at the front lines of the never-ending campaign to eradicate hearing loss. They were not merely selling a product: they were promoters of "a hearing re-education program, with the hearing aid as one of the tools necessary to get the job done."[13] A subgroup of dealers, the "white coat boys," even perceived their authority as akin to that of medical experts. They argued that their years of experience in the field enabled them to diagnose instances of faulty hearing and prescribe treatments, thus fulfilling similar duties to the otologist. In this vein, Sonotone rebranded its dealers as "consultants," eschewing the association of dealers with market capitalism, and instead locating them strictly in the medical profession. For prospects, the blurred boundaries between dealers, otologists, hospital clinics, and, later, audiologists, only served to overwhelm them with buying options and make them—especially the elderly—vulnerable to high-pressure upsells. Scores of devices were available on the market, advertised as powerful and discreet, and varying in price and quality.

The commercial availability of the transistor, which was developed by Bell Laboratories in the 1940s, made it possible to design more powerful monopaks, and then later, one-piece units worn completely in the ears (the behind-the-ear style). Sonotone's 1010 model, released in 1952 and sold for $229.50, was the first transistorized consumer product released in the United States, composed of two vacuum tubes and one transistor made by the Germanium Products Company.[14] Raytheon also began manufacturing and selling junction transistors under license from Bell Labs to hearing aid firms. Like the Sonotone 1010, these new aids were hybrids, but by 1953, Maico Company introduced the first all-transistor hearing aid. That same year, 300,000 hearing aids were sold, a third of which were all-transistor models. Overwhelming demand led Raytheon to become the largest manufacturer of transistors.[15] Transistor hearing aids became so popular that a year later, 97 percent of all hearing aids on the market were transistor models, rendering the vacuum tube obsolete.

The transistor hearing aids, however, were not necessarily cheaper, more reliable, or always practical. They were, however, smaller. Small enough to be contained in spectacle frames, small enough to be worn behind the ear and concealed by clothing or skin color. Small enough so that clothing noise was dispensed with and extra battery packs were unnecessary. And for hearing aid firms, advertising these models required capitalizing on the notion that a particular lifestyle was being sold, one of "better hearing" through "hearing happiness" so that "NOBODY KNOWS I'M DEAF." Correcting hearing impairment became a matter of public service. Firms like Sonotone and Zenith used educational films and magazines to emphasize how "uncorrected" impairments of hearing prevented the deaf from enjoying their lives, drawing parallels to the acceptance of wearing spectacles to correct poor vision.[16] Celebrities became spokespersons for hearing aid firms, while fashion magazines showed female deaf readers how to conceal their aids through glamor to overcome the "ordeal of silence." Dealers sold an ideal of living that was perfectly measured by a prospect's quest for happiness—a wife regaining a husband's affection, a grandmother included in family conversation, a business executive no longer being singled out, or a child raising a hand to answer a teacher's question.

At the same time, this approach served to entrench broader prejudices against hearing-impaired people who did not buy into this vision, namely those who chose to be deaf or culturally Deaf instead. Even the design of the transistorized eyeglasses-hearing aid disseminated stigma against deafness. The first commercial model emerged in 1955, when Otarion released their Model L10 "Listener," with thick frames housing the electronics, and a silicone earpiece to be inserted into the ears.[17] Models from competitors followed in the 1950s: Acousticon's Model A-230 in 1955, Radioear's "Lady America," Beltone's "Hear-and-See" and Sonotone's Model 400.[18] Since the hearing aid aspect of these eyeglasses could be smartly concealed—behind the ear and covered by hair, advertisements for these models emphasized the "invisibility" of deafness. Yet the spectacles feature (whether useful or cosmetic) was *not* perceived as disabling—which was ironic given that industry experts seeking to regulate hearing aids frequently compared them to eyeglasses.

Such stigma was aggravated by revelations that most manufacturers exaggerated the benefits of their models by using misleading advertising slogans to confuse consumers. Before transistor hearing aids dominated the market, there were no major technological innovations to set apart different models. Advertising was thus crucial for companies to demonstrate their advantage over their competitors.[19] As E. F. McDonald Jr., president of Zenith Radio, elaborated, this method was "a planned 'cloak of mystery'—purposely thrown around these instruments. It is made *very* difficult for the average purchaser to judge values."[20] Beginning in the 1950s, the American Medical Association's Council on Physical Medicine, the Federal Trade Commission, and the Better Business Bureau aimed to regulate the marketing and advertising of the hearing aid industry and the vicious competition among industry leaders. Every person with impaired hearing, whether they used a hearing aid or not, needed to be a critical consumer to assess the truth of advertisements, an issue that would intensify during the 1960s with the increased consumer activism. For the consumer, seeking out a hearing aid was further complicated by the fact that different experts—dealers, audiologists, otologists—offered contradictory advice while asserting that aids were *necessary* for fulfilling their obligations as citizens.

Modern Miracles

A family sits in front of their television set watching a variety show featuring a tuxedo-clad trumpeter. Mom sits on the arm of the chair, thin and perfectly coiffed; Dad is in the chair wearing black horn-rimmed eyeglasses and smoking a pipe. Sonny sits on the floor, wearing a striped t-shirt and clutching a teddy bear. This is the ideal 1951 image of a white, middle-class family enjoying their evening leisure time. Except for the fact Mom cannot hear as well as Dad and Sonny, so she sits and pretends she, too, is enjoying the entertainment. She continues pretending, until one day, she receives the "gift of hearing" in the form of a Zenith Royal Hearing Aid, at the low cost of $75. It completely transforms her life, bringing her "from a world of half-heard sounds to the full enjoyment of family activities." She no longer has to pretend. And there are thousands like her.

This was the "American dream" for the deaf, as conveyed by advertisers during the 1940s and 1950s, shaping cultural ideas about what it meant to be a hearing-impaired person. They spread the mes-

FIGURE 5.2. Advertisement for Zenith's $75 Royal hearing aid (1951). © LG Electronics USA. Courtesy of John W. Hartman Center for Sales, Advertising & Marketing History, David M. Rubenstein Rare Book & Manuscript Library, Duke University.

sage that a hearing aid could deliver transformation: the deafened woman, once a neglected damsel, now becomes a crusader for all to become hearing—while maintaining her family and femininity. Moreover, despite disclaimers urging prospects to consult a physician to inquire whether a hearing aid would be useful for them, the implicit message in these advertisements was that not wearing an aid was a failure of self.

Because, as historian Roland Marchand states, advertisements faithfully reflected the values and preoccupations of advertisers, advertising agents, and copyrighters—rather than those of their customers—these messages presented deafness as an affliction requiring a technological fix. For consumers, however, the advertisements (if only indirectly) contained images that they hoped could mirror their own lives. A mother frustrated because her deafness interfered with her relationship with her husband and children could certainly have perceived a Zenith advertisement as aspirational. Plus, with Zenith's inexpensive model, becoming "hearing" was likely a realistic and affordable option for her self-improvement.

Hearing aid advertisements did more than create a fantasy to be sold: they mirrored cultural expectations for the deaf to become hearing citizens. During World War II, for instance, hearing aid firms marketed their products by addressing the anxieties of war and encouraging deaf persons to fulfil their patriotic duty by working on the home front. While most manufacturers ceased hearing aid production to turn their factories and materials over to the war effort, the larger firms, including Sonotone, Acousticon, Zenith, and Western Electric, continued to manufacture hearing aids, in part to fulfil governmental obligations to the soldiers with war-damaged hearing.[21] Zenith, for instance, had designed a new model with subminiature tubes in 1942, but the product was shelved following stoppage of all commercial production. When wartime shortage of manpower became acute, Zenith's president, Eugene McDonald Jr., argued that hard of hearing people could serve as a domestic civilian workforce, especially in factories; many, however, could not afford the high costs of hearing aids.[22] Once granted permission to build a high-quality, ready-to-wear portable unit, Zenith released its pre-war Radionic Hearing Aid, the first cheap aid, onto the market in 1943. It cost $40 (approximately $580 in 2018 dollars); other models sold for upwards of $200.

"You owe it to your Uncle Sam!" Zenith's advertisements for the Radionic declared. "He needs manpower—every available person. A hearing deficiency may keep you out of the armed forces . . . but you can do your fighting on the home front. . . . A good hearing aid enables you to go all out in the war effort."[23] And with a bonus, the aid came with a *"Liberal guarantee."* The demand for cheaper aids made Zenith's Radionic production the largest in the world.[24] Zenith even promoted its hearing aids in radio broadcasts, acknowledging how American free enterprise made it possible for the company to produce cheaper models to achieve its goal for deaf persons to fully participate in civic and social duties.[25] Other Zenith advertisements during the war years addressed the Radionic's "miraculous" ability to bring out attractiveness in a woman and restore her personality with the use of their "Neutral-Color Earphone and Cord" and choices of "Lustrous Ebony" or "Pastel" for harmonizing with her dress colors—but reminding the customer, as always, of the state of the nation: "Buy More War Bonds!" is stamped on the corner of the advertisement.[26]

Sonotone took out full-page advertisements in *Life* magazine illustrating how wartime necessity transformed the country's understanding of hearing loss. The national call for patriotic duty meant that there was "no excuse for delay" in compensating for hearing loss. With a Sonotone aid, over 90 percent of the hearing-impaired could receive "uninterrupted hearing." A 1943 advertisement, for instance, has a private kneeling besides a pilot and his airplane, checking off a list on a clipboard, the earpiece of his hearing aid clearly visible. This scene at an airplane hangar is certainly a noisy one, but deafness was not going to stop Sgt. John M.H. from doing what was best for his country; with a Sonotone, he received two promotions and worked as an instructor in a flying field. Another advertisement details three men in front of a shipyard site: a manager, a marine

FIGURE 5.3. "Deaf Can Fight, Too!" Acousticon advertisement, 1940s.

electrician, and a Sonotone consultant. All three are wearing Sonotone hearing aids, working towards "America's drive for victory," like
scores of other Sonotone wearers in bustling shipyards, aircraft, and
munitions plants. "Just a few years ago," the advertisement explains,
"a hearing aid was a distinct handicap in getting a job. Today, the example of such industrial leaders . . . has upset that old prejudice. A
man wearing a Sonotone is welcome in any plant in America if he
can do the job."[27]

The emphasis on patriotism continued into the postwar period. Even as hearing aid technologies continued to advance with the application of military innovations to civilian products, advertisements emphasized that it was the deaf person's duty to contribute to the strength of the economy. As historian Mara Mills has argued, the development of the Solo-pak hearing aid, which was the first to use the printed circuit, transformed deafness into a variable for military aural rehabilitation experiments, contributing to more intricate hearing aid designs.[28] Acousticon's "Privat-Ear," released in 1959, was designed for men and women seeking to lead "a full and active business and social life."[29] So too, was the "Atomic Age Miracle," which brought "perfect hearing" to all important domains of social, financial, and domestic life.[30] Yet some civilians felt that despite improvements in hearing aid technologies, they were not receiving the same technological advancements with their commercial units as hearing-impaired soldiers were.[31]

While advertisements portrayed aspirational social realities, dealers sold hearing aids. They sought out prospects to directly convey the message of good hearing and provided demonstrations.[32] Some relied on mailing lists, while others, like the American Ear Phone Company of New York, encouraged customers to make "easy commission" by selling at least one Audi-Ear to a friend or neighbor.[33] Indeed, even some dealers were hearing impaired themselves, using their own stories to advocate the benefits of hearing aids, or were disabled veterans who received government-sponsored aids and set up a postwar hearing aid business.[34] Most dealers used direct sales in the prospects' own homes, offering them a private opportunity to understand the degree of their hearing loss and navigate their purchase options. The joy of a family's reaction to the prospect's being able to hear was often enough for the dealer to close a sale. When in doubt, he used precision machines like the Selex-a-Phone, the first Master Hearing Aid—a test with a microphone and receiver to set the approximate gain and frequency best suited for a person's hearing loss to demonstrate the performance and quality of the aids. In the home, the deafened person became hearing again.

For some hearing aid users, their dealer was more than a sales-man: he was the link between impaired and good hearing, a constant reminder of the importance of patriotism and civil duties within the normal postwar culture. He was there to address employment chal-lenges, to provide connections to lipreading services or to otologists. Seldon Klein of Cleveland, Ohio, for instance, "lost his grip and be-came depressed" when his hearing began to deteriorate. After he lost a large portion of his multi-million-dollar fortune, his family re-located him to Fitchburg, Massachusetts. Believing Klein's troubles were largely psychological but unsure where to look for help, they contacted a Sonotone representative in Fitchburg, hoping he could offer suggestions.[35] The representative forwarded the request to Fred W. Kranz, one of the Sonotone Corporation vice presidents, who then wrote to Dr. Hallowell Davis, director of research at the Central Institute for the Deaf in St. Louis. Davis recommended two medical practitioners, one a hard of hearing physician with "broad experience and good human understanding," and the other a psy-chiatrist who worked at Dashon General Hospital with the "aurally handicapped."[36]

It was quite usual for hearing aid users to form lasting relation-ships with their dealers long after the initial sale. American archae-ologist Theresa Goell, for instance, wrote to her Sonotone dealer, William Röwlönd, while stationed at a site in Nemrud Dagh in Tur-key. While a student at Radcliffe College, Goell discovered that her hearing was starting to fade; a diagnosis of otosclerosis meant her hearing would progressively worsen, which she dealt with by lip-reading and wearing bone conduction hearing aids.[37] In June 1955, while stationed at the American Embassy in Ankara, she received a bill for batteries and noticed a curious charge of $50 to keep the purchase guarantee. "I also paid this charge last year," she wrote to Röwlönd, "but when I brought in my three machines on return from Turkey in April and asked for a new one because all three were out of order, I did not receive them and they were repaired."[38]

A few weeks later, the hearing aids were acting up again. "I there-fore see no purpose in paying to keep in guarantee instruments for which you make no substitute for and for which you offer only twenty dollars should I decide to get a new one on my return." Frustrated, Goell phoned her dealer but received no reply. Letters from 1958 indicate that she received improved aids and was happier with S. F. Morgan, her new Sonotone dealer.[39] She frequently kept Morgan updated with how her aids fared during her long voyage and on excavation sites, especially the batteries, which were often unsatisfactory. Battery failure did not appear to bother Goell much; she was known for feigning deafness whenever her batteries ran down, so she could resolve disagreements by postponing further debate until she was able to hear again.

Fashionable Flattery

As historian Anna G. Creadick argues, normality functioned as a powerful measure of American life, even becoming a "substitute for morality" in some instances, but always a goal to strive for.[40] Postwar advertisements regarding hearing normality defined gender roles as depicted through advertisers' social mirror: if wearing a hearing aid was embarrassing, uncomfortable, or an unpopular choice, they strategized it was better to sell *hearing*, not aids, by directly offering a solution to cultural perceptions of unattractive gendered behavior of deafened people. Hearing aids could return women to the dinner and card tables; they gave men effective command of work and family situations.[41]

Advertisements were placed in national magazines such as *Life*, *Ladies' Home Journal*, and *Popular Science* to inspire consumers to purchase and wear hearing aids. In addition, firms published magazines depicting stories of successful hearing-impaired people—"real" people—who overcame their handicap. These stories carefully crafted an image of normalcy as equated to success in work, family, and leisure. Sonotone's educational department released the first issue of *Better Living* in August 1937, with a brightly colored cover illustration of a schoolhouse atop a field, and children at play supervised by two adults. Its message is optimistic: the joy of children's laughter is contagious and hopeful.⁴² *Better Living* sought "to do its part in a great program of public education by means of which there is every reason to believe that the unhappiness, maladjustments, and trials that have accompanied hearing loss in the past can be largely eliminated." Most articles were written by experts rather than seasoned journalists, including by Vern Oliver Knudsen, professor of acoustical physics, and George Skinner, colonel of the U.S. Medical Corps. The magazine's circulation ran into hundreds of thousands; it provided content covering the latest scientific information about the physiology of the ear and improvements in surgical procedures, as well as advancements in hearing aid technology.

As the magazine's name implies, Sonotone recognized hearing problems to be "as broad as life itself—that the uncorrected impairment of a sense as vital as hearing often deprives the individual of the ability to enjoy many of the things that make life worth living." The magazine ceased production during the war years, returning in 1950 with a cover featuring an elegant couple in evening wear; a lapel hearing aid is pinned on the woman's gown. Glamor became the cornerstone of Sonotone's message. Though the company's war advertisements implied that hearing impairment did not excuse citizens from their duty to participate in the war effort, *Better Living* eased the transition to daily living by depicting how easy it was for people to succeed in work and life. Ignoring discussions of war trauma (except the psychological impact of deafness and isolation), articles focused on depicting postwar America, one in which the man went off to work in a suit and tie, while the woman in an apron affixed to her New Look dress and holding a child waved goodbye. In cultivating patriotism by promoting a particular version of middle-class America, Sonotone used its products to link gender with normality.⁴³

Zenith also published its own in-house magazine. First released in 1958 as part of the company's public service mandate, and with a print run of a million copies, *Better Hearing* was free to all who wrote to request a subscription. Offering insight into various problems faced by the hard of hearing, *Better Hearing*, like *Better Living*,

FIGURE 5.4 The cover of Sonotone's *Better Living* magazine, July 1950. Central Institute for the Deaf collection, Series 3, Bernard Becker Medical Library, Washington University School of Medicine.

followed the motto of "Information and Inspiration." Readers were encouraged to educate themselves in the latest developments in medical science and hearing aid research and to turn to people like Kathleen Delahunty, a deaf-blind graduate of Hunter College, for inspiration. Attempting to provide a balanced and complete approach

SMART "Sono-Charm" JEWELED PINS
Cover the "Movable Ear" – the Outside Microphone

Sono-Charms are attractive costume jewelry, designed to slide easily but securely over the "Movable Ear."

The Sono-Charm, with its out-of-sight cord, converts the "Movable Ear" microphone into what appears to be merely a handsome, decorative pin. And the cost is so moderate

that many women will want several to blend with, or to accent, different costumes for party, street or evening wear.

Plain clips are also available, so that if you have a favorite pin, brooch or insignia, you may have it attached by your local jeweler.

Remember, ONLY Sonotone gives you the outside microphone
with no OUTSIDE telltale, dangling wire!

FIGURE 5.5. In the 1950s, Sonotone released their Sono-Charm, jewelry that could disguise wires and microphones of hearing aids. Fashioned with intricate design, they were promoted as a necessary accessory for hearing-impaired women. Advertisement, 1930s. Central Institute for the Deaf collection, Series 3, Bernard Becker Medical Library, Washington University School of Medicine.

to the range of lived experiences of hearing loss and deafness, *Better Hearing* also featured deaf schools, including St. Joseph's School for the Deaf in the Bronx and the Clarke School for the Deaf in Northampton, Massachusetts. Nationally renowned otologists, such as Aram Glorig and Samuel Rosen, were also recruited to write articles.

Both *Better Hearing* and *Better Living* promoted highly gendered messages that were also present in Zenith and Sonotone's hearing aid advertisements. During the 1940s, *Better Living* bestowed its monthly "Man of Distinction" title upon the man with the most inspirational story and echoed Sonotone's advertisement campaign of self-improvement. The February 1952 title, for instance, was given to Sloan McCrea, a Miami-based businessman and civic leader who "helped create America's flourishing resort city." A Sonotone wearer since 1941, when he lost his hearing from a double attack of measles and pneumonia while serving in the U.S. Army, McCrea traded his role as sergeant in the Quartermaster Corps for a routine life as a business executive. McCrea's story is one of overcoming, but also one of loyalty. "Wherever I go," he says, "Sonotone takes care of me." His success, as much as his masculinity, incorporates postwar cultural ideals about what it meant to be a "normal" American male: a leader, a business executive, a powerful figure and a devoted family man, ideals that were embodied in the gendering of hearing aid designs and advertisements as well.

Sonotone's postwar advertising campaigns particularly emphasized gender norms in both the design and the marketing of their hearing aids. One of their free brochures—which was also advertised in *Better Living*—outlined how women could style their hearing aids "for flattery."[44] Bright pink and titled FASHION: *Your Passport to Poise*, the brochure was aimed at the young, fashionable and stylish hearing-impaired woman who wanted to maintain her glamorous self without compromising the functionality of her hearing aids. This was a woman who cared about her visible self and thus did not want to become a "conversation-exile." She would rather be alert, relaxed, and happy with her friends and family than left in isolation. If she was a mother, she was told not to feel sorry for herself and needed to remove herself from the "withdrawal period" that was characteristic of those who were gradually losing their hearing and trying to hide it. She was not one to deceive herself about the reality of her hearing loss, because after all, she was deaf, but not sorry.[45]

The brochure included black-and-white photographs of happy, white, and beautiful women in slim silhouette, offering stylish ways for concealment. A woman could wear a simple scarf knotted at the neck—even with a rose through it—to hide the unsightly hearing aid cords. Her hair was her best strategy; short, flattering hairdos, in stylish cuts, could hide the "button" receiver, which was designed to be "Out-A-Sight." Another option was to wear her "SonoComb" with its "double-duty magic"—it was both a hearing convenience and a beauty aid—discreetly placing the hearing aid into her hair, a nod to nineteenth-century aurolese ear trumpets. Good grooming and guile could be enhanced with the "SonoCharms," inexpensive

FIGURE 5.6. Page from Sonotone's pamphlet, *FASHION: Your Passport to Poise* (1950), advising hearing-impaired women on how to disguise their hearing aids and wires with strategic placing of jewelry, hair, and clothing. Such images reinforced gender stereotypes and cultural expectations for women to maintain a polished and poised image for themselves without drawing attention to their deafness. Central Institute for the Deaf Collection, Series 3, Bernard Becker Medical Library, Washington University School of Medicine.

**No woman
is prettier
than the expression
on her face...**

Glamour cannot substitute for poise or charm. Flawless features cannot disguise this straining, wrinkled brow . . . these troubled eyes . . . this mouth about to say: "I didn't quite hear you." Yet Betty believes that since she doesn't wear any hearing aid, nobody knows she's deaf. Why her very expression shouts it to the world! Finally her husband, weary of shouting at her, convinces Betty that she is deceiving only herself.

How clearly this sweet face mirrors inner poise. Betty doesn't only look different— she is different. Her happy personality reflects itself in the way she sits and walks, the way she listens and talks. And Betty's Sonotone is more than a hearing aid. It's her magic wand to better living! With so many clever new accessories that keep a Sonotone strictly private, Betty can honestly say . . .

"Now, nobody knows I'm deaf!"

FIGURE 5.7. Sonotone's "Nobody Knows I'm Deaf!" campaign emphasized that wearing aids were less conspicuous than the so-called "deaf face" of mishearing. Page from Sonotone pamphlet, *FASHION: Your Passport to Poise* (1950). Central Institute for the Deaf Collection, Series 3, Bernard Becker Medical Library, Washington University School of Medicine.

BETTER HEARING
styled for
MEN...

Every man wants to look his best. Most men who need hearing aids are interested in wearing them as inconspicuously as possible. Sonotone has done several things to make this possible.

HOW TO WEAR YOUR INSTRUMENT

Sonotone makes the instrument or transmitter so small that wearing it involves little difficulty in concealment. Many men wear their instrument suspended by a neckstrap under their shirt, fairly high up, where a four-in-hand necktie aids in the concealment. Others prefer to carry it in the vest or shirt pocket. When worn in the latter, try clipping it to a piece of stiff cardboard, slightly narrower than the pocket and slightly higher. This will keep it steady and will prevent pocket sag, while lifting the instrument high for better reception and easy adjustment. Recent models may also be worn in the trouser watch pocket, or even inside the trousers, clipped to the waistband.

Sonotone also makes available a special belt attachment for small instruments. With it, a hearing aid is carried in a place convenient for adjustment and where it is out of the way as well as out of sight.

FIGURE 5.8. Companies provided men with strategies for concealing their hearing aids in suits. From: Sonotone 1950 pamphlet, *Better Hearing Styled for Men*. Central Institute for the Deaf Collection, Series 3, Bernard Becker Medical Library, Washington University School of Medicine.

costume jewelry with gleaming gold loops, art deco designs, and moonstones, that held the "Moveable Ear" microphone. Matching earrings were also available.

These strategies were intended to capitalize directly on beauty and sartorial expectations for women to be stylish, slim, well-groomed, and docile without showing their deaf face. Cord wires interrupted her silhouette: ugly lines running up her neck or in front of her chest, a visible reminder of her defect. The infernal racket caused by clothing rubbing on the cords was a constant frustration to her attempts at self-improvement through better hearing.[46] The need to hide wires (or even battery packs, before monopak aids and transistor aids were available) posed a challenge for most users. For one thing, the wire could be too short to extend underneath clothing, or too long and become uncomfortable against the skin. Wiring oneself up was itself another challenge: the 1930s and 1940s, vacuum tube hearing aids offered harnesses for securing the aids in undergarments, but most users complained about the discomfort. Acousticon, for instance, recommended women strap their battery packs on their thighs and place the aid between their breasts. Battery packs were heavy and thus were uncomfortable to be strapped on legs all day long—more so when trousers became fashionable for women and the bulging exposure was unavoidable.

Users often made their own adjustments to harnesses or dress to accommodate their aids. England's National Institute for the Deaf, for instance, encouraged readers of *The Silent World*, the institute's in-house magazine that was distributed to American readers, to write in their tips for improving experiences with "deaf-aids." One reader remarked that when she wore her hearing aid, the "lack of freedom was very noticeable." Her full range of movements restricted by the wire and cords, she crafted her own solution by making a pocket to fit the aid and securing it by webbing straps around the neck and under the arms. This way, the device could be worn under a blouse or dress without being exposed. The battery leads were brought down the top of a skirt, and the reader recommended that "if a costume or jacket with pockets is worn it is a good plan to make a small slit at the back of each of the pockets and buttonhole round them." With the lead from the headpiece secured by a hairclip and passed under the blouse, the only part of the hearing aid visible was two inches of wire at the back of the neck—even this could be hidden if the hair was worn fairly long. "The extra patience needed while rigging oneself up," the reader noted, "is well repaid by the sense of freedom."[47]

Gladys Reed of St. Albans offered advice for carrying a hearing aid. The "special handbag container" that came with her hearing aid to be slung over her shoulder, with the earpiece strung inside her sleeve, was a nuisance. To remedy this, Reed made a belt with a pocket on either hip to store the microphone and batteries to be worn under her dress. "This was a great improvement," she explained, "but made my dresses bulge rather. I often sighed for the numerous pockets a man possesses." Eventually she improved on her initial design by making a "small pocket in the center of [her] 'bra,' on the underside, to hold the microphone tightly." She stored the batteries in a small cardboard case inside a linen bag measured and cut to fit the case exactly, but slightly longer to be fastened on her suspender belt and secured by a large safety pin. Like that of the reader above, this method allowed Reed to pass the earpiece wire under her arm and around the back of her neck to reduce its visibility. The setup only added five minutes to her dress time, but at least it "stays put" all day.[48]

Hearing aid adverts certainly made invisibility seem easy. "She's wearing it!" exclaimed a 1951 Acousticon advert, showing a Grace Kelly look-alike dressed in an evening gown, holding a cup of tea, and smiling as her mirror reflection shows there were no hidden cords.[49] "Can you see the Hearing Aid this girl is wearing? Positively not, even she can't see it!"[50] Sonotone offered new hearing aid models with a "slimmer silhouette": white women with cinched waists hiding slim aids in bathing suits, Sunday best jackets, off-the-shoulder evening gowns, and halter-neck tops. Happy, elegant women photographed demonstrating their stylish tricks by using pearl necklaces, brooches, undergarment harnesses to hide wires to increase their confidence and selfhood.

The central message emitted from these advertisements, and especially from Sonotone's fashion brochure, was that "glamour cannot substitute for poise or charm. Flawless features cannot disguise this straining, wrinkled brow . . . these troubled eyes." No illusion or trick could conceal obvious markers of deafness, so it was crucial for women to wear hearing aids; to stop being deaf, and instead, to keep their deafness a secret.[51] As *The Silent World* letters reveal, however, the struggle to disguise wires and ensure relative comfort while wearing the aids was a challenge for most users. Despite newer hearing aid models designed for assisting in dress—aids like Tonemaster's cordless "Barrette" aimed for "freedom that has never before been enjoyed by the hard of hearing woman," or Soundfinder's "Ear Ring type" built to resemble earrings—the reality of dressing with aids frustrated users who expected them to be fully inconspicuous.[52] Vanity, as industry leaders knew, was a driving force for hearing aid acceptance, especially for women.

While women had barrettes and jeweled designs, men had pocket-sized aids smaller than a deck of cards, and aids hidden in tie pins, pens, legion lapels, and cuff links. Acousticon offered a "Super-X-Ear" hearing aid disguised as a watch; it corrected hearing, freed the user from obnoxious clothing noise, and eliminated "most of the objections [that] the hard-of-hearing have against conventional hearing aid designs."[53] Again, clothing noise challenged the acceptance of hearing aids. One user described how he eliminated clothing noise by wearing his instrument under his lapel, with "two button holes for this purpose on each suit so spaced that the instrument's clip goes in the upper hole, and the wires through the lower one. The instrument is thus held solidly, out of sight and in a location where there is nothing to rub against it."[54] The Red Cross also offered the hearing impaired a flannel case that somewhat reduced clothing rub.[55] Just as women made adjustments to dress, men too took it upon themselves to modify their hearing aids to improve comfort. After all, as advertisements indicated, hearing aids were tools for predominantly white-collar jobs, necessities for postwar

FIGURE 5.9. Pages from *I Can Hear Again!*, an Acousticon pamphlet (c. 1930s) highlighting how the problems of deafness could be alleviated with a hearing aid. Messages like this reinforced the cultural stigma against hearing impairment by indicating hearing was necessary for happiness. Kenneth Berger Museum and Archives.

men to succeed by incorporating ideals of business power and masculinity.

These ideals are especially pertinent in one Acousticon pamphlet titled *I Can Hear Again!* A two-page spread illustrates the unlimited benefits a good hearing aid can provide: participation in conversations, hearing a choir or an orchestra, and improved mental clarity. Music and speech clearly define the parameters of good hearing. On the left page, headlined "What Price Deafness?," various scenarios are highlighted, scenes that a hearing-impaired man with a strained deaf face faces daily. The scenes are akin to a bad dream: being socially handicapped, feeling insecure in his business, strained by conversations between himself and others, experiencing loneliness, and being disturbed by people shouting at him. The circular markings on the page, meant to represent the transmission of sound, are darkest by his ear, expanding towards his environment. The facing page shows the opposite. The man is now smiling radiantly, looking at the viewer. The circular representations of sound are lighter, brightest near his ear. "What Happiness Hearing?" headlines the page: family and social life become joys once again, as do companionship, church, and business security.

Highly gendered and designed to be inconspicuous in dress, modern hearing aids were, as industry leaders asserted, a response to their customers' repeated complaints that hearing aids were generally too conspicuous and they'd rather not wear one. Pretending the hearing loss was not severe or completely denying it only served to do the hearing-impaired person an injustice. As one writer said, "It's natural to take pride in our appearance. However, this perfectly, normal, self-respecting human vanity may double-cross us if we don't watch out."[56] Indeed, vanity was such an obstacle that the hearing aid firm Paravox revised its advertising strategy to avoid the use of words such as "deaf" or "Hard of Hearing" after discovering test groups found them objectionable. Instead, Paravox referred to "a limited 'usual HEARzone,' which is a more pleasant way to express it."[57] Advertisements depicted nervous, irritable men and women with an imaginary bubble surrounding their faces—their "usual HEARzone"—which marked the distance beyond which ordinary conversation became unintelligible and difficult to understand. Wearing a Paravox hearing aid, the ads suggested, was an efficient and economical tool for expanding one's "usual HEARzone."[58]

Cloak of Mystery

Hearing aid firms were aware of how the psychological aspects of hearing loss were closely associated with the loss of identity, especially for customers who were late-deafened or lost their hearing from an accident or illness in adulthood. Coming to terms with their deafness required these individuals to reevaluate how to live in a world of increasingly muffled sounds before finding themselves

FIGURE 5.10. Paravox booklet featuring "usual HEARzones": imaginary bubbles to demonstrate the limitations of hearing loss. The ad completely avoids the word "deafness," "deaf," or "hearing-impaired," which tested negatively in marketing focus groups. Central Institute for the Deaf Collection, Series 3, Bernard Becker Medical Library, Washington University School of Medicine.

FIGURE 5.11. Acousticon dealers would carry laminated cards like these to demonstrate the dramatic benefits of wearing a hearing aid. These before-and-after photos certainly resonated with prospects deliberating whether a hearing aid would expose their deafness. From Acousticon dealer's book, *How Hearing Affects the Mind* (1949). Kenneth Berger Museum and Archives.

lost in what one firm referred to as "the valley of perpetual silence." Acousticon, for instance, trained their dealers to begin sales by first understanding the basic psychology of prospects, in that "fundamentally they are handicapped people."[59] The deep-seated insecurity known to the deafened not only made them susceptible to domination by others, but also prevented them from seeking out help when required.

This peculiar characteristic of prospects tended to make them vulnerable to exploitation by unscrupulous high-pressure dealers wanting to maximize their commissions. To set itself apart, Acousticon required its dealers—known as Acousticians—to acknowledge that "this psychological fact can and must be used by you *constructively*. You must use it in the same way a competent, able physician does. You must exercise the same kind of professional domain in . . .

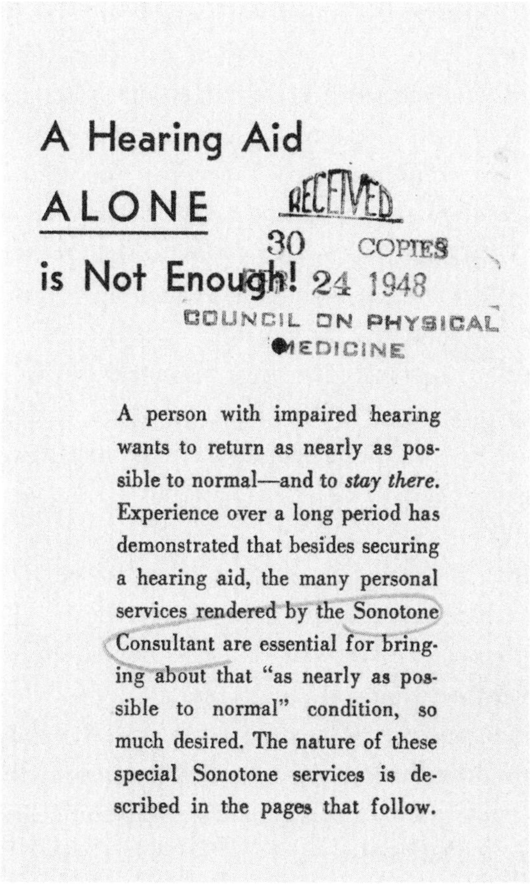

A Hearing Aid

ALONE   RECEIVED

<u>　　　　　</u>   30   COPIES

is Not Enough! 24 1948

COUNCIL ON PHYSICAL
MEDICINE

A person with impaired hearing
wants to return as nearly as possible to normal—and to *stay there*.
Experience over a long period has
demonstrated that besides securing
a hearing aid, the many personal
services rendered by the Sonotone
Consultant are essential for bringing about that "as nearly as possible to normal" condition, so
much desired. The nature of these
special Sonotone services is described in the pages that follow.

FIGURE 5.12. Sonotone pamphlet outlining that additional care was required by their
consultants, for "A Hearing Aid <u>ALONE</u> Is Not Enough!" Pamphlet, 1948. Central Institute
for the Deaf Collection, Series 3, Bernard Becker Medical Library, Washington University
School of Medicine.

the Acousticon Sale."[60] Acousticians were expected to be elite salesmen, held to a higher standard than their competitors, for they were
trained in the rudimentary aspects of psychology, aural anatomy,
and audiology.

Yet no matter how well dealers were trained, they were *not* physicians, even if some worked in a paramedical capacity. This issue
particularly irked otologists, who believed that dealers, if not advertising firms, were misrepresenting the value of hearing aids and
fittings. Otologist Edward P. Fowler Sr., for instance, charged that

Sonotone's references to "individual fittings" and "life long service" could mislead people. "Do you not think it is allocating to yourself," he wrote to Irving Schachtel of Sonotone, "the principles and teachings of the otologists throughout the years?" The company's advertisements included phrases such as "certified Sonotone Consultants scrupulously follow the standard practices accepted by the medical profession"—an approach Fowler argued had no meaning and was misleading and unethical, as it implied Sonotone's consultants were qualified to make medical diagnoses.[61]

The question of ethical advertising plagued the hearing aid industry during the 1950s. E. F. McDonald Jr. noted that while the industry offered a "means of attaining happiness and normal living," too many manufacturers relied on misleading advertising slogans to confuse consumers. The American Medical Association's Council of Physical Medicine agreed. Created in 1925 to combat the dangers of commercialized "machine therapy," the council concentrated its efforts on educating physicians on the value of electrotherapy apparatuses and physical therapy equipment, heretofore a fad exploited by manufacturers and salesmen.[62] After World War II, the council formed various committees to oversee different sections on physical medicine and rehabilitation, including a committee on hearing aids and audiometers. Leading otologists and scientists tested hearing aids and ranked them based on performance, quality, and workmanship, passing on their recommendations for the council's list of approved hearing aids that met both technical standards and the guidelines set by the Federal Trade Commission's Fair Trade Practice Codes. A select few of the committee members, including Hallowell Davis, also analyzed hearing aid advertisements to determine whether a manufacturer and model deserved the council's seal of acceptance.[63]

As well, the National Better Business Bureau published a list of hearing aid advertising recommendations that were developed through a voluntary cooperative program with industry manufacturers to build public confidence in hearing aids. The recommendations were not legally binding but rather were presented as supplementary principles to the trade practice rules for the industry as promulgated by the FTC in 1944.[64] False or "bait" advertising was to be avoided, as were exaggerated claims and scare tactics that unfairly exploited the consumer's desire to minimize or eliminate the visibility of their devices. Any advertisements creating a false impression of "precision fitting" or implying medical analogies—the showroom as a "clinic," the salesman as a "doctor"—were deceptive. Echoing the recommendations by the council's Advisory Committee on Audiometers and Hearing Aids, hearing aid advertisements were forbidden to use unqualified superlatives—"the finest," "the best," "the world's smallest"—unless supported by refutable, competent proof.[65] These recommendations were modified throughout the 1950s to better encapsulate changes in advertising standards; the advisory committee, for instance, acknowledged that even if an instrument met high standards of technical excellence, problematic advertising and misrepresentation undermined the integrity of the product and resulted its place on the "unacceptable" list.[66]

Even local chapters of the American Society for the Hard of Hearing formed their own Committees on Advertising Standards. They directed that no hearing aids would be advertised in its monthly newsletter that were not on the Council on Physical Therapy's accepted list. Misleading exaggerations were not permissible. Chapters could reserve the right to reject or alter advertising for their newsletters that did not conform to standards.[67] The Chicago Society for the Hard of Hearing, in particular, addressed the importance of truthful advertising, for the more informed its members were, the better they would be able to help "people come out of their shells and to go back into a hearing world."

As council secretary Howard A. Carter noted, "If manufacturers are to prosper in business they must sell hearing aids and encourage those who need hearing aids to buy them"; misleading methods, on the other hand, degraded the industry and solidified consumer resistance to buying what they perceived to be a nuisance.[68] The advertising of hearing aids, Carter argued, required copywriters to avoid slogans such as "hear naturally," "perfect hearing," or "the world's smallest." As Carter stated in one report, "Vacolite and Beltone cannot both be right. Each one claims the smallest hearing aid. In fact, Vacolite's dogmatic statements are something of a masterpiece. Beltone is on the ragged edge when it hides the cord behind the main instrument in its photograph and claims 'This is all you wear.'"[69] Any promotion of "invisible" hearing aids or claims of "hidden hearing" were particularly problematic, for they promoted an unhealthy attitude towards hearing aids.

Indeed, Acousticon, a subsidiary of Dictograph Products Company, came under fire in 1949 when the advisory committee recommended that its new "Hearette" model be rejected from the council's approved list of hearing aids.[70] The problematic advertisement boasted that the "amazing new development" required "nothing to wear, no receiver in the ear, no cords, no bulky batteries," with small-print text containing what Davis described as a "come on": a strategy to draw in a consumer with dazzling promises, only to upsell a more expensive model instead. "I believe the Fair Trade Practice makes 'come on' illegal," Davis wrote to Carter in February 1949, "if the item advertised is not actually available for sale and is not as represented; but I don't see how we can stop this."[71] By June, Dictograph Products sent out a memo to its Acousticon distributors revealing that the Hearette advertising was discontinued; two months later, the company delivered a forty-page report urging the council to reconsider its removal of the Acousticon Hearette from its list of approved hearing aids.[72] The pleas failed; the 1950 list does not have any approved Acousticon or Dictograph products.

Another industry strategy to sell hearing aids was to convey the fact that its workers and salesmen were also hearing impaired. Dealers often revealed their own hearing impairment during direct sales, finding that empathy enabled hesitant prospects to feel understood. Sonotone frequently illustrated users at work and at home; other companies used celebrity endorsements to convey similar messages. Reformer, activist, and best-selling American author Dorothy Canfield Fisher became the first celebrity spokesperson for hearing aids in 1938. Deafened at the age of fourteen, she never acknowledged her impairment, refusing to discuss or even write

about it; speech-reading served as her primary means of communication, since for her deafness was "a hinderance in every way."[73]
Though Fisher may not have accepted her hearing loss, readers of
Better Living perceived her as a "success story," proving the benefit
of celebrity testimonials. Zenith's advertisements featured famous
deafened Americans, including Eleanor Roosevelt, Charles Edison,
and Rupert Hughes who, despite being able to afford any type of
aid at any price, preferred the $75 Zenith model. In 1949, Paravox
rolled out a new campaign with celebrity spokespeople to promote
their products.[74] Even if the featured celebrity did not wear hearing
aids (or was not deaf), as long as they knew the importance of "good
hearing," and were connected with the motion picture business or
radio, it was enough to deliver Paravox's message. Bob Hope's approved statement—"It is gratifying to know that with Paravox Hearing Aids the hard of hearing, everywhere, may now enjoy my Radio
and Movie Picture Shows"—emphasizes the *possibility* that the hard
of hearing *may* hear with Paravox hearing aids; it does not promise
hearing. Nor is it a testimonial or endorsement, but rather "a frank
statement by a man who is really interested in the listening audience
because his livelihood depends on listeners."[75]

Hearing aid advertisements increased in intensity during the "bitter commercial war" among manufacturers when Zenith released
its $75.00 Royal Hearing Aid with a full money-back guarantee.[76]
Schachtel, for instance, accused Zenith of engaging in deceptive
trade practices and insisted that the council was unfairly reprimanding his company despite its public commitment to ethical and honest
advertising. If any advertisements for Sonotone hearing aids were
found to be downright dishonest and "aimed at hurry-up sales," he
argued, it was not the fault of the company, but of hearing aid clinics and unscrupulous dealers who placed ads in local newspapers.[77]
Leland A. Watson, president of Maico, used the same defense.[78] A
follow-up inquiry by the council confirmed that Zenith refunded
the money without question, but attacks on Zenith continued.[79]
W. N. Brown, general manager of Zenith, wrote to council member
Ralph E. DeForest, explaining that since Zenith launched its cru-

MRS. ELEANOR ROOSEVELT

World-famous wife and mother; Senior United States Representative of the United Nations General Assembly; author, radio and television commentator; internationally respected and admired for her interest in, and understanding of, all peoples.

HONORABLE CHARLES EDISON

Son of the late Thomas A. Edison; former Assistant Secretary and then Secretary of the Navy; former Governor of New Jersey; guiding force as officer and/or director in many nationally known civic, educational and industrial organizations.

MR. RUPERT HUGHES

Author, playwright, producer, poet, biographer, composer; chief assistant editor of the 25-volume History of the World published by Encyclopaedia Britannica; veteran of two world wars; Hollywood writer, Doctor of Letters, director and commentator.

These three great

Americans can afford any

type of hearing aid

at any price. They wear

the seventy-five dollar

Zenith hearing aid.

BIOGRAPHICAL DATA BASED ON "WHO'S WHO IN AMERICA."

FIGURE 5.13. Eleanor Roosevelt, Rupert Hughes, and Charles Edison were listed by Zenith as their celebrity "great Americans" who wore the company's $75 hearing aid. © LG Electronics USA. Courtesy of Central Institute for the Deaf Collection, Series 3, Bernard Becker Medical Library, Washington University School of Medicine.

sade to lower the cost of hearing aids in 1943, they had "been continuously subjected to vicious, loose, and dishonest attacks" coming
from competitors objecting to its crusade. He continued:

> Competitors have complained about us to the Department of Justice,
> to the Better Business Bureau, and even to your American Medical
> Association. Some have even gone so far as to urge certain suppliers
> of component parts of hearing aids to refuse to sell their products
> to Zenith. All of these attacks have failed for the simple reason that
> Zenith is doing exactly what it set out to do, namely selling to the
> public a hearing aid of the highest quality at a price far below that
> charged by most competitors.[80]

Opposition to Zenith's crusade concentrated primarily on the
impossibility of manufacturing an affordable high-caliber hearing
aid that could fit all degrees of hearing loss, as Zenith claimed to
do. Either the low-cost aids were cheaply made or they were a publicity stunt to increase wartime sales. In defense, McDonald stated:
"I think the time is coming in the not so distant future when the
racket will be taken completely out of hearing aids and there will be
no hearing aids selling at over $100. . . . The biggest handicap that
Zenith has had is its low price because the people have been educated to pay higher prices and they believe that in order to get the
best they have to pay the higher price."[81]

The battle against Zenith intensified after the company published a pamphlet titled *Frauds and Facts* to educate its customers
about the manipulative nature of the hearing aid industry. Drawing
out copy taglines from competitors' advertisements, the pamphlet
aimed to clarify misstatements in order to encourage sales of the
company's $75.00 Royal Hearing Aid and to defend its technical in-
tegrity against the claims of competitors. Indeed, *Frauds and Facts*
focused on the fact that most advertisements for hearing aids did
not list prices, or that if prices were listed, the direct sales approach
usually favored upselling. To test this, Zenith apparently sent out a
hard of hearing individual to four local dealers to purchase an aid at

the listed price of $69.50, offering cash payment if the dealer also provided the ten-day money-back guarantee. All four dealers, the investigator reported, convinced him that the cheaper hearing aid was not designed for his hearing loss, and pushed models priced between $199.50 and $229.50—a justification of course, for Zenith's $75 hearing aid, powerfully and rhetorically displayed next to three leading competitive models, their chassis removed to expose their material parts. Unsurprisingly, the publication of *Frauds and Facts* angered other hearing aid companies, who charged the booklet was nothing more than Zenith's transparent attempt at competitive copy.[82] Competitive adverts insisted that hearing aids must be "tailor-made," not "ready-made," and that just as "one hat won't fit ALL heads," one hearing aid won't help all people to hear, no matter the price. Yet competitors too attempted to manufacture their own "cheap model," determining whether it was possible to devise one for $15 or $20, even though $20 was roughly the cost of component parts alone.[83]

Paying through the Ear

Though the AMA Council on Physical Medicine, working in conjunction with the FTC and the Better Business Bureau, aimed to regulate the marketing and advertising of the hearing aid industry, this viciousness among top competitors played out until the 1980s. Hearing aid advertisements were not simply reflections of cultural ideas and values: they served as a form of communication that had traceable effects in the world. The issue of fair trade practices and ethical advertising intensified in 1962, when a public health survey reported that hearing loss affected more Americans than any other chronic condition. Approximately 8.5 million Americans had some degree of hearing impairment; those aged sixty-five and over were the most affected group.[84] The report revealed that 52.9 percent of hearing aid users past the age of sixty-five never received an audiometric examination prior to their hearing aid purchase. The elderly, who were the most vulnerable to unscrupulous salesmen, required

taking you "back of the scenes"

there has not been any basic change in hearing aid design or principle in the last five years... With that in mind—consider this—

here is a Zenith and three leading competitive models...

  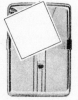

| Zenith $75 | Set A $199.50 | Set B $199 | Set C $185 |

THEY'RE ALL ABOUT THE SAME SIZE—SAME MATERIAL—AND MUCH ALIKE IN LOOKS.

now let's look at what is inside—**here are unretouched photographs of the same instruments opened for inspection—**

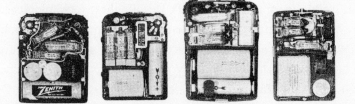

comparison shows they all use approximately the same material and parts and require about the same amount of labor to assemble.

form your own conclusions

FIGURE 5.14. Page from Zenith's pamphlet *Frauds and Facts* depicting what the hearing aid industry referred to as "competitive advertising": malicious attacks on competitors by juxtaposing Zenith's $75 Royal Hearing Aid next to more expensive models. © LG Electronics USA. Courtesy of Warsaw Collection of Business America, Lemelson Center, National Museum of American History.

additional counseling to help them navigate the hearing aid market and any testing or services required.

The problem of hearing aids, then, was a consumer problem. Reports indicated salesmen were on the prowl to seek out the elderly. They carried their own testing devices with them to make on-the-spot sales, often pressuring customers to accept devices without any proper audiometric test or fitting; sometimes they insisted on a sale even if the elderly customer did not require a hearing aid. One eighty-year-old man testified that a salesman conned his wife out of $600. "They make all kinds of promises," he continued, "and keep none of them once they get their hands on the money which they demand in advance."[85] Citizen complaints suggested that dealers not only pressured the elderly for sales, but even confused or misled them, inflating prices to sell ill-fitting or poorly constructed hearing aids. Not all dealers followed suit, though. Many manufacturers and dealers offered honest, helpful service, improving performance, reducing the size of hearing aids, and developing new techniques to improve fitting for their customers. Despite these technological and service improvements, customer complaints persisted, launching broader concerns about trade practices and the general structure of the hearing aid industry.

On 18 and 19 July 1968, the U.S. Senate Subcommittee on Consumer Interests of the Elderly, chaired by Senator Frank Church, held hearings to discuss the hearing aid industry. As Church observed, the meeting was to discuss a fundamental question: "What more should be done in this Nation to help older Americans—those most vulnerable to deafness and near-deafness—to save them from the isolation, demoralization, and hazards that occur when hearing deterioration becomes severe?" The key issue was cost: with over three hundred hearing aid models on the market ranging in cost from $100 to $400, and limited assistance for examinations, hearing aids for the elderly appeared beyond their reach.

One crucial aspect of the congressional testimony was the inadequate involvement of the medical profession with the hearing aid industry—or at the very least, the failure of hearing-impaired persons to seek out proper medical care. Deafened people assumed they could purchase an appropriate hearing aid as readily as a pair of spectacles, without any professional advice. Dealers convinced prospects otherwise, insisting that specialized, intimate care was the only avenue for better hearing, for hearing aids were meant to be personally fitted, which dealers insisted they could do better than medical experts. Despite this insistence, as Dr. Eldon L. Eagles testified, the lack of proper medical attention was a major reason for consumer dissatisfaction and eventual abandonment of hearing aids. As indicated by the National Health Survey, up to 35 percent of persons with binaural hearing loss had never been tested by an otologist.

"Just as we no longer buy spectacles on a basis of trying on a few pairs until we feel that we notice some improvement," Eagles stated, "neither do we regard deafness as a simple mechanical situation, correctable by nothing more than a simple mechanical procedure."[86] Another problem was the lack of standardization for audiometers (specifically their improper calibration), which resulted in poor reading of hearing acuity and thus misdiagnosis or misfitting of hearing aid requirements. Dr. Joseph L. Stewart, consultant for the speech pathology and audiology department of the National Center for Chronic Disease Control, emphasized that hearing loss was primarily a public health problem within the allied domain of otolaryngologists and audiologists, even though hearing aid distribution was a commercial venture dominated by crafty salesmen. These salesmen, he argued, were ill-suited to properly determine the type, extent, and duration of hearing loss as measured by the audiometer, or to properly assess hearing aid selection for clients—especially given the extraordinary technological gains made in the industry.

John J. Kojis, president of Maico Hearing Aid Instruments and former president of the Hearing Aid Industry Conference, explained that on average, it took customers five years after realizing they were losing their hearing to seek technological assistance. "We discovered an unmistakable truth in our field—that most people who are becoming deaf will not initiate a program to get help."[87] Seeking out medical assistance and advice was further complicated by the fact that otolaryngologists and audiologists in the 1960s competed with hearing aid dealers for access to consumer-patients, thus confusing hearing-impaired persons about their choices for proper audiological care. Of approximately 400,000 people who purchased hearing aids, more than 95 percent bought them from dealers, risking poor fitting and poor acoustic gain. Other experts at the hearings insisted that the best approach to reforming the hearing aid industry was through education—to educate consumers about their choices and to address public awareness of the benefits to be enjoyed from proper hearing aids. Otologists were crucial for convincing any patient "of the value of a hearing aid, which is frequently rejected as stigma," especially patients with conductive deafness, who greatly benefited from the device.[88] Additionally, otolaryngologists and audiologists needed to play a much greater role in the hearing aid industry, and national organizations (such as the National Hearing Aid Society) needed to better work with dealers and manufacturers.

These congressional hearings were characteristic of the rise of consumer activism and watchdog groups in the 1960s. The problems of the hearing aid industry—soaring prices, unfair competition, misleading information, preying salesmanship—indicated that the $132-million-a-year industry, as outlined by the FTC, was in dire need of reform. And again, the need for reform centered on the role of the medical expert within the capitalist marketplace: how could the otolaryngologist legitimatize their profession when it was continuously being undermined by hearing aid dealers who professed to have greater training and insight into the nature of hearing aid fittings? Backed by the FTC, audiologists argued that dealers with no medical training were misleading patients by exaggerating their audiogram charts. Otolaryngologists and the AMA told dealers to limit themselves to fitting and leave their medical clinical aspirations out of their profession.

By the 1970s, a new age of consumer activism emerged in response to this chaotic market. Hearing aid distribution tactics, such as high-pressure door-to-door sales, cold-calling, and misleading pamphlets, increasingly caught the gaze of watchdog groups and government agencies that produced reports advocating for regulatory changes in the industry. One study was launched by the Retired Professional Action Group (RPAG), an organization funded by Public Citizen, Inc., the advocacy group founded by American political activist Ralph Nader. In 1968, Nader recruited a team of bright law, medical, and engineering students to lead investigative probes aimed at holding corporate powers accountable and igniting public interest advocacy. Dubbed "Nader's Raiders," the group achieved notoriety with a groundbreaking investigation into the FTC, discovering that despite the institution's mandate to protect American citizens from dangerous products and false advertising, collusion, corruption, and incompetence among its officers failed to detect violations or even enforce what limited powers they had. "The Nader approach," explains anthropologist Roger Sanjek, "combined a populist critique of corporate power with muckraking research, local activism, and legislative reform."[89] The FTC investigation led to the agency's overhaul and reform, and Nader leveraged his popularity to set up a cottage industry to publish investigative reports and establish a number of advocacy groups to engage in activism for consumer rights, including the RPAG, a group that enlisted retirees to work on social issues.

In 1972, Nader hired Elma L. Griesel to oversee the RPAG and launch a sixteen-month investigation of the hearing aid industry. Over the course of the study, RPAG staff associates in Washington, D.C., combined with the Gray Panthers activist group in New York City and volunteers in four states, contacted more than a thousand persons—hard of hearing individuals, their families, ear specialists, audiologists, dealers, manufacturers, social workers for the deaf, state and federal rehabilitation offices, consumer groups and agencies, state offices of the attorney general, Medicaid, and dealer licensing boards—to lay out the scope of the industry. To obtain firsthand information about the practices of hearing aid dealers and consumer protection, the study began with eight Baltimore volunteers between the ages of sixty-eight and eighty-two years, who were sent to have their hearing evaluated under clinical conditions at the Hearing and Speech Center at Johns Hopkins Hospital, after which they would visit Baltimore hearing aid dealers chosen at random by RPAG from the telephone directory. The results revealed that in 42% of the visits, dealers recommended hearing aids to customers whose clinical evaluations indicated that they did not need them, and that all dealers insisted it was not necessary to visit an ear specialist or physician prior to deciding on a hearing aid. The volunteers also discovered the prices of aids varied greatly and that dealers offered no explanation of why certain models were more expensive than others. Furthermore, dealers not only pressured customers to buy the costly versions, they also falsely claimed that hearing aids could prevent further hearing deterioration—if not fully restored normal hearing—in their attempts to satisfy disillusioned customers.

The RPAG study documented the monopolistic character of the hearing aid industry, in which four companies accounted for half of all sales and every major manufacturer was cited by the FTC for anticompetitive practices.[90] Even ear specialists did not go unscathed in the study. They were labeled as "ostriches" for not working hard enough to protect the rights of their patients and for failing to regulate hearing aid dispensing, especially for the elderly. The study called the FTC "deaf" to its need to regulate the quality and advertising of hearing aids. And despite its numerous citations against manufacturers, the FTC was nothing more than a "watchdog with a soft heart." The study concluded by stating that medical examinations should be mandatory for prescribing hearing aids, home sales should be banned, and that a regulated distribution system modeled on the Veterans Administration was necessary to overhaul the industry. After all, of 15,000 dealers in the industry, only 2,114 received the minimal twenty-week home study course offered by the industry's trade association, and most received no licensing or testing competency before entering the retail market.

Titled *Paying through the Ear*, the three-hundred-page report was controversial prior to its release by Public Citizen in September 1973. The RPAG's (unpaid) director, retired diplomat Joseph E. Wiedenmayer, resigned in February, stating that the report "seems to confirm what I was afraid of. It would seem that such a project could—inadvertently or otherwise—only confuse or frighten the hard of hearing." In a telephone interview with the *Albine Reporter-News*, Wiedenmayer claimed that his resignation stemmed from a disagreement "with the philosophy and approach of the Nader organization."[91] In May, the National Hearing Aid Society and the Hearing Aid Industry Conference denounced the report, distributing a "fact sheet" to all congressional offices and sending out press releases charging that the RPAG's allegations were "not only factually wrong but materially harmful."[92]

The report was disseminated widely. In Iowa and Massachusetts, it sparked further state hearings, and the federal Department of Health, Education, and Welfare launched a task force to study the hearing aid industry. Even as the industry came under intense scrutiny, it continued to challenge *Paying through the Ear* for promoting "premeditated slander and fraud."[93] The Joint Committee on Hearing and Hearing Aids expressed its fear that multitudes— "perhaps millions—of hearing-impaired people will be driven by the Nader report even further back in their reluctance to get help they desperately need."[94] The committee insisted that the RPAG report was "wholly inaccurate in some of its analysis of industry operations," focusing on the few negative instances and thus grossly misleading the public about the state of the industry. They labeled the Nader report a fraud and called the RPAG's research "amateurish."[95] "The dealer," said Samuel Lybarger, president of Radioear, "has been under attack by some of the most malicious propaganda the industry has ever seen."[96] Other industry spokesmen asserted that such reports were nothing more than "scare stories" serving as overdramatization of rare exceptions in the industry.[97]

Wasting Time

As a result of the RPAG report, the FTC cracked down on trade practice rules for the hearing aid industry, even reaching out to experts like Hallowell Davis for advice with tricky language in adverts.[98]

What differentiated dealers from other sellers (such as opticians), however, was the sophistication of their own products, which relied upon medical and audiometric diagnoses; given the fragile place of expertise within the hearing aid industry, dealers needed to be regulated. The 1950s rules—prohibiting general misrepresentation, bait advertising, unfair trade practices, and deception—needed to be modified to account for the role of dealers. Oregon was the first to pass a hearing aid licensing law for dealers in 1959, but testing and prescription remained by and large unregulated in other states until the 1980s. The question of manufacturing and retail costs and defining the parameters for the roles of experts continued in another special hearing before the Senate Subcommittee on the Consumer Interests of the Elderly on 11 September 1973.[99] Dr. Robert Ruben, for instance, testified that the "first obstacle in the way of our citizens obtaining a proper hearing aid is the excessively high cost"; the cost of manufacture was approximately $15–$35, but the retail cost was $200–$600.[100] Further, hearing aids continued to be sold without consumers receiving a proper audiological examination.

Four years after the hearings, the FTC released a report concluding that "profit-hungry salesmen" often deceived customers (especially the elderly) into buying expensive hearing aids that were of little benefit. Beltone, for instance, encouraged its dealers to meet their selling quota by selling "lemon aids" (hearing aids that do not work properly or are overpriced). Further, some dealers even misinterpreted hearing test results, recommended the wrong type of hearing aid, or promoted outdated models.[101] The FTC recommended that prospective buyers refrain from purchasing hearing aids without a doctor's recommendation, or at least to waive their right to one in writing. After ten years of toiling to protect customers from defective hearing aids, in 1984 the FTC stated that the evidence compiled during the 1970s and earlier was now out of date. Commissioner George W. Doylar argued that a new $14,000 survey was required to update the evidence and properly assess the state of the industry; otherwise the FTC would be wasting everyone's time.[102]

"Wasting time," reported Michael Isikoff of the *Washington Post,*

"—and a lot of other things—is precisely what some observers think the FTC has being doing for years with the hearing aid case. As some see it, the story of the hearing aid rule is a classic example of how a federal regulatory agency can devote enormous resources and years of analysis to a problem without ever reaching a conclusion."[103] Furthermore, the hearings revealed that the public greatly misunderstood the nature of hearing loss, especially regarding whether certain types of hearing loss could be dealt with surgically or technologically. Any advancement in hearing aid miniaturization would be impractical if there were no consumers willing to purchase the product; though pricing and advertising of hearing aids were reminiscent of fraudulent quack cures, the aids themselves were useful products for amplifying sound for deafened people to hear.

The congressional sessions also revealed the need to address the ignorance surrounding the belief that hearing aids provide an instantaneous restoration of hearing and that the more expensive an aid, the more capable it was of accomplishing that. As Nanette Fabray of the National Association of Hearing and Speech Agencies stated, "It is not like putting on a pair of glasses and having your vision improved. It takes time to learn how to wear a hearing aid and it takes time to get used to wearing a hearing aid."[104] These claims would resonate towards the closing decades of the twentieth century, especially as medical experts testified modern hearing aids are "both effective and reliable."[105]

Beyond Eyes of Incredulity

You, as parents, have a real interest in all that pertains to the problem of
deafness. You want your child to overcome his handicap so that it ceases
to be a handicap. Sometimes you dream a little. Often you may hope too
much. Again, you may expect too much. You search for remedies. You
would spend your last dollar for a remedy or cure. There is nothing you
would not do for your deaf child. That is only natural. Before you seek out
the miraculous, it is well to deal with the present because today is so important. Each day is important in the life of a deaf child.

Leonard M. Elstad, 1948[1]

To refer to the implant as the bionic ear was to draw attention away from
its technical effects and towards its symbolic ones. It was no longer audiograms that were evoked, but the Six Million Dollar Man with his superhuman capabilities.

Stuart Blume, 2010[2]

It's an iconic photo.

The face of a young child who was born deaf, hearing sounds for
the first time. Jack Bradley, a photojournalist from the *Peoria Journal
Star*, captured the exact moment a doctor fitted five-year-old Harold
Whittles with an earpiece and switched it on. Originally printed in
the February 1974 edition of *Reader's Digest* in an article titled "Un-
forgettable Moments Caught on Film," decades later Bradley's photo
achieved notoriety on Internet "best of" lists. The photo is "shocking," "miraculous," "unbelievable," "influential," "heartbreaking,"
"heart-warming," "amazing," and "evocative." It has been circulated

thousands of times on social media sites, where commentators candidly express their reactions: the photo brings tears, it reminds us of our humanity, it tells people to "count their blessings," it evokes personal experiences with deafness and hearing, and it triggers debates about language, culture, and technological achievements. For many people, the photo immortalizes more than Harold's incredulity. It serves as a testament to the wonders of scientific medicine and the abundance of hope. It is a glimpse into the future.

At the same moment Harold was hearing his first sound, a monumental shift was occurring in the scientific community: a new technological solution posed the potential to completely eradicate deafness. But unlike other deafness cures, this one precipitated accusations of cultural genocide. Medicine's unflagging determination to vanquish diseases became perceived as an invasive trespassing upon deaf bodies rather than an altruistic endeavor.

Just a year before the publication of Bradley's photo, scientists, physicians, and policy makers gathered in San Francisco at the First International Conference on Electric Stimulation of the Acoustic Nerve as a Treatment for Profound Sensorineural Deafness in Man. Prompted by a research group at the University of California, San Francisco, which had been conducting auditory experiments on animals, the conference focused on the ramifications of implanting electrode technology into the brains of deaf patients for the purposes of restoring their hearing. Experiments to electrically evoke hearing originated in the 1800s with Alessandro Volta's insertion of a fifty-volt battery-connected metal probe into his ear canals that, once switched on, caused a "jolt in the head" and sounds resembling "a kind of crackling, jerking, or bubbling."[3] Though Volta declared his experiment too dangerous, further experiments during the nineteenth and twentieth centuries encouraged the notion that the auditory nerves could be directly stimulated with an electrode inserted in the brain. In 1957, two French surgeons, André Djourno and Charles Eyriès, performed the first direct electric stimulation on the human auditory system, but their patient could not afterward understand speech properly. By 1961, the "tinkering" experiments of the 1950s evolved into the first single-wire electrode developed by Dr. William F. House and successfully implanted in a deaf patient. Over the next ten years, the implant was refined to improve the viability of wide-band speech signals and perceived pitch variety required for sufficient sound transmission and speech recognition. Despite criticism by otologists and neuroscientists, in 1974, the result of nearly 175 years of research was finally introduced into medical literature and to the public: the cochlear implant.[4]

In most cases of deafness and severe hearing loss, there is damage to the sensory hair cells in the cochlea. The cochlea is the center of hearing—the O'Hare Airport, as Lydia Denworth succinctly illustrates, where sounds arrive, change form, and travel out again in a complex pattern of activity.[5] Hearing aids are thus ineffective in these cases, as they only amplify sounds to transmit to the cochlea; without hair cells, sound signals cannot be processed to reach the auditory nerve of the brain. A cochlear implant (CI) bypasses the damaged part of the cochlea to deliver sound in the form of an electrical signal directly to the auditory nerve to stimulate hearing. The device consists of two parts: an external portion resembling a behind-the-ear (BTE) hearing aid that includes a microphone, a speech processor, and a transmitter converting signals from the processor to electrode impulses; and an electrode array implanted under the skin that collects the impulses and sends them to the auditory nerve.

The first commercial CI, the House/3M—named after House, who developed the first clinical CI with engineer Jack Urban—was released in 1973 and used a single-channel electrode, intended to aid lipreading. Once introduced for widespread use, hundreds of deaf adults and parents of deaf children rushed to undergo the procedure. Over a thousand people were implanted from 1972 to the 1980s, until the 3M Company sold their rights to the Cochlear Corporation in 1989. Within this time span, House's single-channel system was criticized for failing to adequately replicate speech signals for users to discriminate intelligible speech. Research proved that multielectrode CIs contained speech processing systems better able to overcome biocompatibility issues. Though improvements to CIs occurred rapidly, the technology was highly controversial since its introduction. Skeptical auditory neurophysiologists and otologists questioned the device's long-term feasibility. Some called it a cruel hoax, another in a long line of failed deafness cures manipulating desperate people's desire to hear. Almost right from the onset, members of the Deaf community resisted the implant, especially for children who could not provide informed consent. In 1977, the U.S. National Institutes of Health commissioned a study evaluating the performance of CIs, revealing that the devices, on average, provided a better quality of life for users as communication aids useful for voice modulation and speech production. The study—often called the "Bilger Report" after research director Dr. Robert C. Bilger—did much to dispel the skepticism and scientific controversy surrounding CIs.

After nearly twenty-five years of experimentation, the Food and Drug Administration approved the House/3M device for widespread commercial use only in adults, though some children had been implanted before 1984. The agency's deputy commissioner, Mark Novitch, reflected on the profound implications of the technology: "For the first time a device can, to a degree, replace an organ of the human senses."[6]

CIs, however, do not replicate the natural sense of sound. If anything, they are tools to help the deaf person navigate their surroundings and discern between different sound sources while learning speech. Their early promotion as "bionic ears" alluded to their cyborgian ability to transcend boundaries between reality and fiction, appearing at once both contemporary and futuristic. They were "tiny computers" that promised to bring sound to 70 percent of deaf people.[7] They were a "functional replacement of the ear," engineered for the deaf to hear everyday sounds, from the roar of a jet engine to the crackling of leaves.[8] The promotions were also confusing, using terms such as "electric ear" and "implanted hearing aids" as synonymous with CIs, even though the latter referred to amplification, not restoration of sounds.[9] Following improvements in speech processor technology and multichannel electrodes in CIs, on 27 June 1990 the FDA approved the proposal by Cochlear Corporation to market their CI for surgical insertion in deaf children over the age of two. The headline for *American Health* read: "New Hope for Deaf Children: Implant Gives them Hearing and Speech."[10] Crossing over to mainstream medicine as greater numbers of qualifying patients received the surgery, CIs were heralded as a "medical innovation that breaks through the wall of silence" by providing children the opportunity to lead "a more normal life."[11]

Sold as an effective and necessary treatment, CIs were marketed primarily to hearing parents of deaf children.[12] Even now, hearing parents remain the primary consumers of CIs, not d/Deaf persons; this is unsurprising given that, according to the U.S. National Institute on Deafness and Other Communication Disorders (NIDCD), more than 90 percent of deaf children are born to hearing parents.[13] Despite informative leaflets and detailed explanations from otologists, some parents nevertheless hold unrealistic expectations about CIs, presumably obtained from media coverage of the "bionic ear" and narratives of children overcoming all odds to hear. As Stuart Blume writes in *The Artificial Ear*, his 2010 ethnographic survey of CIs, media accounts nearly always emphasize the hopefulness of the device, playing a vital role in helping to secure resources and patients for surgeons and thus making implantation possible.[14] Yet the CI is *not* a miracle cure for deafness.

The optimistic nature of media accounts tends to ignore a crucial fact: despite improvements to the technology, many deaf individuals do not qualify for a CI. Most cases of implantation are in children born deaf and implanted before language acquisition (between two and five years of age) or adults born with hearing that has gradually diminished through age, genetic disease, or illness. For those implanted with a CI, the "switch-on" is an incredibly difficult transition. Recipients must adjust to an array of ordinary noises they might never have heard before: wind hitting windows, traffic din, or the flow of water from a tap. CIs create a representation of sounds that assists the user to understand speech and sound sources; because of this, speech can sound robotic or filtered as if everyone were speaking underwater. Moreover, users with a CI require extensive therapy to deal with the overwhelming experiences and with adjustments to life, which can be especially difficult for younger children. All these "deaf people hearing for the first time" stories also leave out a crucial—and certainly less glamourous—aspect of implantation. The constant physical pain, adjustment struggles, therapy difficulties, and of course, exorbitant costs associated with implants and surgery, or even the complex medical decisions that go into selecting CIs. The costs of CIs vary according to several factors, including evaluation of hearing loss, the surgical procedure, hospital stays, hardware, and postoperative rehabilitation. The price range for a CI in the United States is $40,000–$100,000 or more. Since 2004, public health care plans, including Medicare, Medicaid, and the Veteran's Administration, cover CIs. While federal law requires all state Medicaid programs to provide coverage for CIs for eligible children under twenty-one years of age, coverage for adults varies by state. Some states provide no coverage at all, while others stipulate that only one CI will be covered, making it difficult for adults with bilateral hearing loss.

Nevertheless, many believe CIs are the ultimate solution to deafness, a valuable technology—if not a "cure"—for restoring a deaf person to "normalcy." Why choose to be deaf, or to communicate in American Sign Language (ASL) or lipreading when there is a tool that could "improve" your life? I've worn hearing aids most of my life and communicate primarily through speech; over the years, I've also learned basic ASL. People feign surprise when they realize that I wear hearing aids, commenting that I "speak well" for a deaf person. Other times, I slip: years of speech training reduced to stutters as my "deaf voice" resurfaces when I repeatedly mangle my words. I've never considered myself a "hearing person" even when I wondered if the latest medical advancements were suitable for me. When I was younger, I referred to myself as "hard of hearing"—wanting, perhaps, to distance myself from classmates who were deafer than I was, and to shed the "deaf" label. In adulthood, I find myself frequently saying that I am deaf. That's the truth, according to my audiogram chart, which measures hearing loss in various registers; anyone with hearing loss at 90 decibels is classified as "profoundly deaf." That's me. Yet I'm still surprised when I complain about being dissatisfied with my hearing aids and people ask me why I haven't considered a CI. I still remember the shock I felt when I told my audiologist that my first digital hearing aids were a failure, and she looked at my audiogram and told me I would qualify for a CI and should consider getting one.

From the outset, the debate over cochlear implants has resided in highly emotive territory, centered on the transforming meaning of deafness, with the pendulum swinging between diversity and impairment. If deafness is a difference, why force a "fix"? Why, for that matter, even consider deafness as a "health issue" at all?[15] Is it kindness or cruelty to open a child's skull and wind wires through the cochlea to the brain? To "rob that child of a birthright of silence"?[16] And, for that matter, how can a CI be marketed as a "cure" when it is a fleeting solution, only activated when the implant is turned on? In the words of medical anthropologist Bharat Jayram Venkat, "What does it mean to be cured when you might need to be cured again, and again?"[17]

In 1993, the National Association for the Deaf (NAD), the premier American organization for Deaf people, condemned CI surgery and declared it a violation of defenseless children's autonomy. Their sentiment is best captured in a cartoon published in *Silent News*, a Rochester-based newspaper for the deaf, and reprinted in the *New York Times*, illustrating a deaf child before and after a CI. Before, he is a happy boy, a "normal" ASL-using member of the Deaf community. After, he is an unhappy child, forced into years of speech therapy to "try" to fit into a "hearing world." In the latter, he has "no chance of being normal and is [literally] scarred for life."[18] The cartoon succinctly captures the reverence for ASL, the touchstone of Deaf culture and identity, but it also explores the ostracism experienced by many CI users in d/Deaf society. Choosing to receive an CI in adulthood could mean being displaced from the community as Deaf culture militants proclaim CIs to be nothing more than weapons of cultural genocide, armed to eradicate ASL.[19]

Thus language was, and is, the center of the controversy surrounding CIs. The nineteenth-century emphasis on the acquisition of spoken English through lipreading and speech, rather than sign, transformed cultural expectations of deafness by placing the measurement of "success" on speech instead of hearing. The technological and medical advancements of the twentieth century further encouraged this emphasis, as product testimonials extolled the virtues of devices that enabled deaf consumers to regain their speech and hearing. This shift, however, was a disaster for the Deaf community. Deaf teachers lost their jobs to hearing teachers and educational attainments for the deaf plummeted owing to poorly developed literary skills. Forced to learn lipreading (or speech-reading), to twist the contours of their tongue and lips to articulate sounds that always appeared foreign to them, deaf people suffered. Learning speech when you cannot hear your own voice is incredibly difficult; focusing your education on the pronunciation of words, rather than the understanding of ideas, hampers your intellect. By the mid-twentieth century, the deaf were more marginalized than they had ever been, forbidden to use the language of their choice after sign language was banned as a form of instruction in deaf schools.[20]

Myriad technological and medical options for improving hearing, however, afforded deaf individuals options for constructing their identity and navigating the hearing world. After decades of activism and scholarship fiercely debating both sides of the CI issue, the NAD finally recognized the diversity within deaf community and within the deaf experience.[21] The organization's 2000 position statement on CIs was a radical shift from its 1993 counterpart: acknowledging that technological advancements had the potential to foster and enhance quality of life (including hearing aids, text telephones, emails, closed captioning, audio loop, and video-interpreting devices), the NAD acknowledged that "cochlear implantation is a technology that represents a tool to be used in some forms of communication, and *not a cure for deafness"* (my emphasis). They do not guarantee cognitive development or reduce reliance on visual language and literacy development. Furthermore, the NAD stated that parents have the right to make informed choices for their deaf and hard of hearing children, but it is important to address the "adverse effects of inflammatory statements about the deaf population," including the pathological view disseminated by the medical community that deaf persons needed to be "fixed" by CIs.

As sociologist Laura Mauldin contends: "The view that deafness is a medical problem is the crux of clinical approaches to deafness and a starting point from which clinical interventions begin."[22] By framing deafness as an impairment rather than a difference, invasive corrections of the ear can be justified. After all, if we willingly accept cataract surgery to restore sight, prosthetics to restore limb function, pacemakers to control heart rhythms, then why are CIs problematic? Isn't it unfair to force deafness as a cultural ideal upon deaf children, especially deaf children born to hearing parents who have no ties to or interests in the Deaf community, as is the case for 90 percent of deaf children? The parental choice whether to raise these children as Deaf is a deeply personal one. Sometimes parents put all their eggs in one basket, hoping that CIs along with a series of extensive speech and aural therapies, and significant time, effort, and financial investment, will make their child embrace hearing. Sometimes the child, deafened since birth, hates hearing, leading to language delay, an inability to communicate, frustration, and behavioral problems.[23] For deafened adults, on the other hand, the CI could be another useful tool to aid their already mapped out deaf/hearing world—not perhaps as a cure, but surely as a technology for adaptation rather than assimilation.

"Cure technology keeps evolving," asserts Eli Clare in *Brilliant Imperfections*, "but the ideology behind it stays the same." The miniaturization of hearing aids, the popularity of faddish treatments, and the powerful agency of surgeons have all become entangled with a narrative of progress, whereby eradicating or concealing bodily differences is a step towards normalization.[24] To what extent does this need to standardize human populations as "progress" become nothing more than "poorly whitewashed eugenics?"[25] Blume argues that our cultural obsession with miracle cures resides in our recognition and faith in medicine's ability "to reduce the risks we run and eliminate obstacles to our living long and healthy lives."[26] We believe in scientific advancements and we hang our hopes on a solution that is around the corner. Even in 1986, when members of the Oral Deaf Adults Section of the Alexander Graham Bell Association for the Deaf and Hard of Hearing gathered in Chicago for their annual convention, they arrived under the banner of the convention's theme: "Making Miracles Happen: New Technologies, New Techniques. See the Future in Chicago."[27] A hundred years of deaf people's desperation to correct their failings, falling for sham psychometrics, charismatic hucksters, and fallible procedures, as they weaved between deafness and national hearingness — and the proposed impact of technologies on the deaf experience continues to fuel discussions and controversy.[28]

Within these ever-shifting boundaries of cure and control, deafness cures are thoroughly embedded in the cultural imagination, deeply anchored in American conceptions of normalcy. Some questionable "cures" might conceivably have had a physiological basis, but they were often reported anecdotally with scant detail that merely distorted the nature of the cure and reduced it to pure conjecture. Advertisements for deafness cures certainly promulgated the obsession with finding a cure, gravitating from mere product descriptions towards a commercialized ideal of happiness. They promoted devices not just to improve hearing, but to improve one's self through an egalitarian ideal of success, while incorporating a consumerist identity within the American way of self-improvement. Even the audacity of medical con men, the lack of legal and medical oversight, and the outright corruption of salesmen did little to discourage the quintessential quest for a cure. By the 1970s, as hearing aids became more powerful and otologists solidified their authority over medical treatments for hearing loss, many questionable cures

FIGURE 6.1. Cover of Radioear advertising pamphlet, *Join the Thousands Who Enjoy Real Hearing Happiness*. © Radioear, Division of Diagnostic Group LCC. Courtesy of Central Institute for the Deaf, Series 3. Bernard Becker Medical Library, Washington University School of Medicine.

died a natural death, weakened by the fact that the momentum generated by popular publicity and advertising could not be sustained indefinitely.[29] In their place, other "cures" took hold: copper bracelets, natural pills, and even spas offering ear candling services. For those unable to afford the exorbitant costs of hearing aids—ranging from $2,000 to $8,000 for *one* aid—cheap disposable versions and "personal sound amplification products" could be purchased online.

During the late 1980s and early 1990s, engineers were developing hearing aids that used microprocessors to manipulate incoming sounds. This new digital technology enabled users to amplify the specific frequencies as needed, a process called equalization control. While analog hearing aids (the predecessor to digital aids) work by amplifying all sounds, digital aids convert sound waves to digital signals, producing an exact duplication of each sound, rather than simply amplifying it. Marketed as a superior technology for the fast-paced digital age, digital hearing aids have, for the most part, replaced their predecessors. It doesn't necessarily mean they are a better technology.

My hearing aid specialists, Chris and Nick, have known me for twenty-five years. My primary school, Dublin Heights, was part of a program arranged with a local hearing aid dealer, Peter Keller, who visited classrooms to check our hearing aids and replace our earmolds. I remember lining up with my classmates, feeling Mr. Keller's hands gently pull my earlobe as he inserted the otoscope into my ear canal to ensure that everything was in "working" order. One year, Mr. Keller arrived with his two trainees—Chris and Nick—who in their youthful eagerness watched as their mentor assessed the children; we were told that Mr. Keller was approaching retirement and eventually his new employees would take over the practice. It was a strange realization to learn that the man whom we had become accustomed to seeing once a month, who reassured us and placed us at ease as we squirmed and giggled when the wet silicone mold was injected into our ears, would be replaced. His visits stopped shortly thereafter, when government cutbacks eliminated the program. From then on, to visit Chris and Nick, I had to go to their

office in downtown Toronto. Established in 1922 as Acousticon of
Toronto, the first hearing aid office the city, the company eventually
specialized in pediatric audiometry and instrument fitting when Mr.
Keller took over in the 1950s. Once consisting only of Mr. Keller and
a secretary, the business expanded to several employees and a visiting audiologist. Chris and Nick have welcomed me warmly over the
years, watching me grow from an awkward preteen, to a university
graduate, and now a PhD and a writer. Mr. Keller, long retired, is still
present in the office: a large painting of his likeness overlooks the
reception area, his bespectacled eyes the most prominent feature.

When digital hearing aids entered the market, despite Chris and
Nick's encouragements, I resisted them. I had tried different analog
hearing models over the years and had finally found a pair that best
suited my needs. The thought of getting a new and improved model
did not appeal to me. I dreaded the exercise of readjusting to my
surroundings with an instrument—figuring out which angle of my
head best amplified sounds through the microphone, how to adjust
my body to block out background noises to concentrate on conversations, and how long I could wear my hearing aids before collaps-
ing from exhaustion. As I went into the office for regular checkups, I
repeatedly declined Chris and Nick's suggestions to try digital hearing aids. Then one day, taking my hearing aid in for a tune-up, Nick
notified me that most hearing aid companies were no longer manufacturing analog hearing aids or their replaceable parts. I was eligible
for funding to cover the cost of new aids, so if I was considering getting a new hearing aid, the time was now: if my analog pair broke,
it would be difficult and possibly more expensive to repair it. "Jai,"
Nick said, looking up from my files, "did you know you're the last of
my patients to convert to a digital aid?" Apparently, I had stood my
ground for ten years.

The new pair were the smallest, sleekest, most inconspicuous
pair I've ever owned. For the first time, I didn't worry about my ears
sticking out or pain from the pressure of wearing an object behind
my ear all day long. They were delicate pieces of beauty that disappeared in my ears and skin, somehow a reification of the hundreds

of 1950s hearing aid advertisements in which I was now engrossed for research. My analog pair appeared comically large in comparison. I had arrived at the digital age, but not without consequences. For one thing, I lost autonomy over my hearing aids—gone were the days of me tinkering with them, taking out my box of tools to fix a regular blockage and avoid a trip to Chris and Nick's office. Instead, to correct fittings, I was hooked up to a strange contraption that connected to Chris's computer as he adjusted my range of sounds. Gone were long telephone conversations with my friends, as I had to choose between a volume-control button or a telecoil switch (which picks up magnetic signals for telephone conversations). Instead, I got months of agonizing migraines as my brain accustomed itself to digital sounds. I could no longer adjust my body to manipulate or block out undesired sounds. Walking down a busy street, I clearly heard conversations down the street and all sorts of incomprehensible noises, but not the voice of my friend next to me. I ended up spending more time without my hearing aids than with them, luxuriating in the self-imposed silence I had spent my childhood wanting to escape. In the silence, it was peaceful. In the silence, there was no pain.

Certainly, twenty-first-century technologies transcend communication barriers by making it easier to get in touch with people, to hide behind screens and phones to avoid awkwardness. It is easier to send a text or an instant message than it is to make a phone call. After years of training—straining—to hear on the telephone, I lost the luxury of phone calls with my first digital hearing aids. Suddenly I felt like I had regressed, especially since all alternative forms of communication—texting, emails, IMs—did not, or could not, replace medical or corporate reliance on the telephone. A giant wall had suddenly separated me from the world, telephone wires taking hold in cracks and crevices, spreading as they scaled the wall. To make calls, I had to ask my partner or my sisters for help, my dependency leading to feelings of infantilization. I had come too far to turn back now. I spent years negotiating my place in the world, to adjust myself to a cacophonous hearing society, and I resented being placed in silence without my permission. This new technology shredded parts of me; it was not until my second pair of digital hearing aids — though bulkier, heavier, and more painful—did I begin to feel like myself again. They were far from glamourous, but they were also more powerful, with multiple settings for me to control the flow and ease of conversations. The telephone is still difficult.

Now we've advanced beyond digital aids and bionic ears.

In 2013, three-year-old Grayson Clamp became the first child in the United States to receive the newly FDA-approved auditory brainstem implant (ABI) at the University of North Carolina. On 21 May, the device was activated and Grayson's reaction to the moment his father said "I love you" is captured on video, the switch-on factor clearly evidenced by the toddler. Created for children born without an auditory nerve or with a damaged one, the ABI bypasses the inner ear entirely to stimulate neurons directly at the brainstem. Like hearing aids and CIs, the ABI is not an instant fix; recipients require extensive therapy and training to process and interpret sounds. Then there's the middle-ear implant (MBI), which transmits sounds to the inner ear through a microphone inserted behind the ear, which then converts sounds to mechanical vibrations; this technology is primarily for individuals unable to wear traditional hearing aids but also unqualified for a CI.

Futuristic technologies bring excitement by transforming the imaginary to the real.[30] Genetic engineering, for instance, which publicly promises the end of human genetic defects, including hereditary hearing loss, realigns the balance between cure and control. In his book *The Politics of Deafness*, Owen Wrigley argues that the "difference" of deafness as a variation from the hearing norm becomes "a focus as part of the attempt to overcome it, the desire to incorporate it into our next level of inclusivity." Through genetic engineering, deafness is perceived as something to be ultimately abolished, either through short-term tactics like assimilation, or through permanent neurosurgical "repairs."[31] Scientists have found that they can trigger the growth of new hair cells in the inner ears of mice; this could potentially repair Usher syndrome, a disorder stemming from a genetic abnormality in which inner-ear hair cells cannot convert sound vibration into electrical signals. On the other end of the genetic engineering spectrum, screening for deafness could be used to preserve a culture and identity, as in the controversial 2002 case of a lesbian couple in Maryland who, in a variation on the "designer baby" narrative, chose to have a deaf baby by obtaining sperm from a donor with five generations of hereditary deafness.[32]

These technological solutions blur the hearing line, as scholar Christopher Krentz argues, for they give "more control over who is on the hearing or deaf side of the line."[33] Where technology fits within these continuously shifting meanings of deafness depends on how the line is drawn and who controls the narrative. There are now wireless hearing aids with Bluetooth capabilities for connecting to other devices, aids with locator apps, rechargeable hearing aids, self-fine-tuning aids to handle different trebles for easier listening. There's an array of digital aids and "hearables" that not only help you to hear, but also enable you to track and monitor your health data. Yet debates surrounding Deaf culture, technological assimilation, and societal normalcy remain ongoing precisely because of a fundamental demarcation: that one group perceives deafness as a disability, and the other sees it as a culture, but the former group holds a disproportionate amount of power. Why are Bluetooth headphones/earbuds/earphones "accepted" without a second glance but not in-the-ear hearing aids (especially since "earphones" used to be a synonym for hearing aids)?

It may be difficult to understand that not every deaf person wants to be "fixed," just as not every deaf person wants to communicate in ASL.[34] The choices of communication and technological adaptation, like the quest for a cure, are deeply personal ones, especially given that for a deaf person, their deafness influences all aspects of their lives, barring one: "They are not deaf when there is no occasion to hear," as deaf adman Earnest Elmo Calkins asserted in 1948.[35]

After months on the road uncovering archival stories of deaf experiences, my partner Geoff and I headed home. Relieved to finally be settled, we looked forward to home-cooked meals and home-bound writing sessions. There were boxes to unpack and research materials to organize. I had thousands of photographs to sort through, hundreds of pages of scattered notes to analyze, and plenty of writing and rewriting to occupy myself during these cold winter days.

One night, as Geoff was working at his desk, I took out one of the boxes. It contained objects from my childhood that I had forgotten about. Black and white photos of my parents and family members. A ceramic bear that my younger sister gifted me for my nineteenth birthday. A photograph of my brother taken the day he was born, housed in a wooden frame I had made in woodshop when I was thirteen. The "Yes I Can" award I received in 1995 from the Council of Exceptional Children for educational and personal achievement. Random knickknacks, birthday cards, and assorted papers. There was also a blue folder with torn edges, my name printed on the cover, with "Carmel School 1987–88" written below it. A wallet-sized photograph was stapled on the top-right corner: I am five years old with rosy cheeks and hair parted in two plaits, wearing a smart white dress and jacket, a small pink flower pinned on the lapel. A soft smile is hinted at in my lips.

Inside the folder are several pages written in Arabic. I deduced these to be my school records and progress reports from Kuwait. I do well for reading, writing, and composition in English; I do fine for reading and composition in Hindi. I passed all my classes, with my teachers noting in English that I "must work hard," and "keep it up!" and that I have an aptitude for drawing. Within the folder are also copies of visa applications for several countries we traveled to when I was a child, and a school transfer certificate acknowledging my withdrawal because I was "leaving for the U.S.A."

A separate green file with more papers had been inserted into the blue folder. The first document I examined was a peach-colored report, written in Arabic, but I could make out a few handwritten scrawls in English. My name, my ethnicity ("Indian"), the name of my school and a dated entry: 19/9/87, meningitis. Another entry confirmed that "she got deafness last year after fever." Pages of reports from the Ministry of Public Health outlining the results of my audiometric test, which measures hearing loss in terms of the degree of impairment. The outcome of the test is recorded on an audiogram, a graph plotting your audible threshold to sounds, as heard on the audiometer; the vertical axis represents volume measured in decibels (dB), with "normal" marked at 0 to –20 decibels. Sounds become louder lower on the graph, with 0 marking the softest sound a person can hear—this is called their hearing threshold. The horizontal axis represents the frequency (pitch) in hertz (Hz). If sounds are louder than 20 dB and you still cannot hear them, then there's a possibility you have mild hearing loss; the lower your hearing is marked on the vertical axis, the worse it is. Over the years, audiograms have marked my hearing loss at between 80 and 94 dB: I am unable to hear most sounds.

One entry in the green file dated December 1986, on which my name is misspelt, includes a handwritten note from the audiologist: "Dear Sir, pure tone Audiometry shows profound deafness Both [*sic*] sides. . . . Bilateral absence of acoustic reflexes maximum intensity of 125 dB for pure tone & white noise. . . . There is no useable hearing." The audiologist recommended fitting for powerful binaural aids and intensive lipreading, voice, and auditory training. "If rehabilitative measures are not adopted," the audiologist warned, "she may loose [*sic*] the existing speech & language in course of time & also loose [*sic*] the vocal quality. She has to rely on lipreading & Auditory training as she is a school going child." The prognosis: "dependent on practice." Another audiogram, taken eight months later by another examiner, confirms my bilateral profound sensorineural hearing loss (SNHL). "Mother noticed some improvement in hearing recently," the audiologist reported, and "child is doing well in school." However, the results of the test were probably "gross responses," as apparently, I was "not too cooperative." A scratched-out comment noted: "could not test—patient did not feel like listening."

I sat on the floor of our living room, surrounded by reports and files from my childhood, and read that last sentence. The words overwhelmed me, washed over my being in an awakening—maybe confirming, finally, what I had denied for years. I had never pretended to be hearing, but I certainly didn't—or couldn't—own up to my deafness. It has been the cause of much anxiety throughout my life, straining relationships and frustrating communication. But it has also been a fundamental part of my identity, something as inseparable from me as the color of my skin or eyes. Reading my childhood audiogram files made me realize how far I had come from that prognosis. So much training and therapy, isolation and loneliness. All this swirled in my mind with the hundreds of stories I had read of other deaf and hearing-impaired people. Suddenly, it became too much.

Deafness is not a tragedy. That is important for me to say. It is not a tragedy. The painful nature of deafness is the contradiction in the "conversion-driven history of deaf people," as Brenda Jo Brueggemann terms it.[36] It is a conflict of impulses to repair and the need to acknowledge diversity, a contradiction in the way normalcy is glamorized when disabled people overcome their imitations, but scorned when they choose to embrace them.[37] "Second in importance only to sight, our hearing is vital to our safety, happiness, and success in life. The person with ears that do not hear is dreadfully handicapped," states a 1950 pamphlet for the John Hancock Mutual Life Insurance Company.[38] Almost seventy years later, the message persists in our culture. Hearing-impaired people, tired of the cacophony drumming into their ears, decline to wear an aid or be fitted for CIs or MBIs and end up stigmatized not so much for being *deaf*, but for refusing to be *hearing*.

"If we give a deaf person hearing," writes Andrew Solomon in his majestic 2012 book, *Far from the Tree*, "are we releasing him into fuller selfhood, or compromising his integrity?"[39] My parents decided to seek out medical solutions not because they perceived me as "broken," but because to them, my deafness was an illness, as is the case for many hearing parents whose child become deaf. My parents were never given the choice about whether to raise me as Deaf, but I'm sure if hearing aids had failed me, they would have. My mom told me once that it was upsetting for her that she couldn't communicate with me. I had shut myself in a world of my own making and no matter what she did, I never let her see or understand how I was feeling. If the *mirchis*, copper bracelets, divine blessings, and hearing aids all failed, she would have kept looking until she found a way to enter my world and communicate with me, if only to tell me that I was loved.

I don't have a photograph or video of myself "hearing for the first time." I have a school portrait of myself at age six, hair in a braid, wearing a pink sweater with a badge with handwritten words on it. The badge was given to me by my kindergarten teacher, the first teacher I had after immigrating to Canada. She had earnest blue eyes and I thought she was the most beautiful person I had ever met. I didn't speak or hear, but she understood me somehow; I knew this because she would calm me down when I enclosed myself in a corner of the classroom, lost in thoughts and tears streaming down my face, pretending to play with the toys in front of me or mindlessly flipping through the pages of a book. Nothing made sense in the classroom. She spoke, but I heard no words. I saw in her eyes what she was trying to say, her patience, and her struggle to make this little girl understand and feel that she belonged. There was no belonging in a classroom full of children who taunted and laughed, pointing fingers and ignoring pleas to play.

She often stroked my head, telling me how much she liked my braid. I turned away from her. The kids were mean, and I felt I had no place in the classroom, that school was no place for me. My mind's reality and the reality I lived never matched up for me, so how could it for her?

Then one day she did the most astounding thing of all.

Every week a child was chosen to be the leader of the classroom: to direct story time or choose what games to play. You had to be smart, popular, and creative to be chosen the leader. Once you were chosen, a "Fearless Leader" badge was bestowed on you, and all the other children became your friends, respected you, wanted to talk to you, and tried to play with you. I wanted that. My heart soared and sunk week after week as I was cast aside as another child gasped and giggled, running up to the front of the classroom to be pinned with this precious jewel. Some got to go up more than once. Each time I had to hold back tears. This was never going to be mine, no one could understand me, so I could never say how much I wanted to have this. But my teacher did. She knew I wanted it, that I had to have it. Perhaps more importantly, she knew it would give me the sense of belonging I was aching for—that I was not alone, that I was not a "freak" or underserving of friendship.

I raised my head one day when I felt her tapping on my shoulder. She smiled so warmly at me that I was momentarily distracted by her beauty and didn't realize that she held in her hands the badge of inclusion for me. That feeling stays with you. The kindness that breaks the dividing line between your mind's reality and the reality of the world. She gave me the badge on Picture Day, giving me a permanent reminder that I was not alone, but empowered.

I take out this photo quite often, sometimes to simply remind myself of the long journey of my deafness, the difficulties I encountered—but more importantly, to feel connected to the distant past, to people who felt the same as I did, who experienced the same stigma, and who, in their own ways and own negotiations for a cure, found themselves.

Acknowledgments

The idea for this book began with my blog From the Hands of Quacks. What initially was a platform for me to document my research and writing struggles evolved into a site chronicling the sweeping history of "deafness cures" and quackery. It is the readers, then, who flock daily to the blog and assorted social media sites who are the first recipients of my gratitude. I am grateful to my editor Karen Merikangas Darling, who took a chance to encourage a rough idea into fruition. I thank the anonymous reviewers and the editorial and production teams at the University of Chicago Press, especially Tristan Bates, Michael Koplow, and Kristen Raddatz.

Research for this book would have been impossible without the expertise and generosity of archivists, curators, and librarians. Philip Skroska at the Bernard Becker Medical Library, Washington University School of Medicine, Arlene Shaner at the New York Academy of Medicine, and Katherine Ott at the National Museum of American History have my immense gratitude for going above and beyond to help bring this project to completion. Special thanks are owed to Douglas Atkins (U.S. National Library of Medicine), Melissa Barton (Yale Collection of American Literature), M. Donald Blaufox (Museum of Historical Medical Artifacts), Hannah Clutterbuck-Cook (Medical Heritage Library), Solenne Coutagne (BIU Santé), Amber Dushman (American Medical Association Archives), Laurie Fink and Rebecca Newberry (Science Museum of Minnesota), Adrian Fisher (Bakken Museum), Lynn T. Frischkorn (American Academy of Otolaryngology–Head and Neck Surgery Library and Archives), Melissa Garfe (Medical History Library at the Harvey Cushing/John Hay Whitney Medical Library at Yale University), Kristina Hampton (Saint Louis Science Center), John Hawks (Kenneth Berger Hearing Aid Museum, Kent State University), Roger Horowitz (Hagley Museum and Library), Rachel Ingold (Rubenstein Rare Books and Manuscripts Library at Duke University), Ivan Kaproth-Joslin (Rochester School of the Deaf Archives), Danielle Kovacks (University of Massachusetts-Amherst Archives), Nick Munagian (McCormick Library of Special Collections at Northwestern University), Mark Murphy (Morris Library, University of Delaware), Alison L. Oswald and Kay Peterson (Archives Center, National Museum of American History, Smithsonian Institution), Rosemary Riess (Special Collections and Archives, David D. Palmer Health Sciences Library), Virginia T. Seymor (Harry Ransom Center, University of Texas at Austin), Christopher Shea (Gallaudet University Library and Archives), Laurie Slater (Phisick Medical Antiques), and H. Dominic Stiles (UCL Ear Institute & Action on Hearing Loss Libraries). Thanks also to the librarians and archivists at the Schlesinger Library, the Huntington Library, the Wellcome Library, the Detroit Public Library, the Detroit Historical Association, Sterling Memorial Library at Yale University, and the Harry Ransom Center at University of Texas at Austin. I am indebted to Carl Erickson for our lovely email correspondence and for sharing materials from his family archives relating to George and Frances Way. Given how much time I spend on Twitter, I am tremendously glad for my global #AcademicTwitter, #dishist, and #histSTM buddies and followers who kept me sane during intense bursts of isolated writing and for answering my late-night questions with such profound solutions.

Research trips would have not been possible without generous financial support from: the Social Sciences and Humanities Research Council of Canada Postdoctoral Fellowship, the Klemperer Fellowship in the History of Medicine at the New York Academy of Medicine, and travel grants to the Lemelson Archives Center to the Rubenstein Rare Books and Manuscripts Library, the Bernard

Becker Medical Library at Washington University, and the Harvey Cushing/John Hay Whitney Medical Library at Yale University.

I am grateful to the following colleagues and friends for their extraordinary service in providing support and feedback at various stages of this book: Jacqueline Antonovich, Geoff Bil, Eve Buckley, Stephen Casper, Jessica Clark, Esme Cleall, Nathaniel Comfort, Rebecca Davis, Dorian Deshauer, Cara Fallon, Lindsey Fitzharris, Michele Friedner, Greg Gbur, Daniel Goldberg, Jeremy Greene, Roger Horowitz, Iain Hutchison, Claire Jones, Stephanie Kerschbaum, Caroline Lieffers, Beth Linker, Elizabeth Neswald, Katherine Ott, Robert Ruben, Daniel Sampson, Karen Sayer, Marion Schmidt, Arlene Shaner, David Suisman, Bharat Venkat, Bess Williamson, and Alun Withey. I am especially grateful to Robert K. Jackler, who invited me to his Stanford office to talk all things quackery; he not only had a printed copy of my dissertation on his coffee table, but graciously let me borrow an unpublished manuscript that was pivotal in shaping my research.

To Graeme Gooday, Coreen McGuire, and Mara Mills—whose respective work on disability technologies intersects with my own—thank you for all the inspiring conversations, for the many meals across the world, for your infallible mentorship, and above all, for your generous friendship.

The audiences at NerdNite Toronto, at the University of Minnesota History of Science and Technology Department, the New York Academy of Medicine, and at the History Department at University of British Columbia provided wonderful feedback for improving chapter drafts. Additionally, I thank the editors and readers of the following sites where extracts of this book were published: *The Atlantic*, The Devil's Tale blog, the Remedia blog, and the *Canadian Medical Association Journal*.

Though much of this book was written in self-imposed exile, it would never have been completed without the incredible support of my family: my mom, Paramjit, my sisters Sundeep and Sukhmani, my brother Jaskaran, and my extended family, David McDougall, Donna Bil, Bryan Bil, Erin Bil, and Zoona Khalid—thank you for being my cheerleaders. I owe a lifetime of gratitude to Kuldip Virdi, Menolly Lysne, Rose Ghaemi, Sarah Kay, Harry Dhesi, Nico Salidas, Erin Kaufman, and my furry companions Lucy, Gizmo, and Lizzie. Wolfie, you are the joy of my life. One day you'll be able to read this book and I hope it makes you proud.

Finally, to my husband, Geoff Bil. This is as much his book as it is mine. No one has spent as many hours as I did, carefully reading and editing every page of this book, nor spent as much time inspecting and sorting through archival records. Thank you for joining me on the road, for cooking dinner so I could write, for taking care of Lizzie, for sharing your intellectual insight, and for your formidable editing skills. Thank you for holding our family together during my illness when it seemed our world was falling apart. I love you, I love you, I love you.

Abbreviations

BAKKEN	Bakken Museum
CANFIELD	Norton Canfield Collection Ms 1293, Manuscripts and Archives, Sterling Memorial Library, Yale University
CID	Central Institute for the Deaf–Max Goldstein Collection, Bernard Becker Medical Library, Washington University School of Medicine
DBC	Dorothy Brett Collection Ms 27, Charles Deering McCormick Library of Special Collections, Northwestern University Library.
DC	Disability Collection ACI319, Lemelson Archives Center, National Museum of American History, Smithsonian Institution
HDP	Hallowell Davis Papers, FC22, Bernard Becker Medical Library, Washington University School of Medicine
HHF-AMA	Historical Health Fraud Collection, American Medical Association
JHC	John W. Hartman Center for Sales, Advertising & Marketing History Collection, Duke University
KBHA	Kenneth Berger Hearing Aid Museum and Archives
WCBA	Warshaw Collection of Business Americana, Lemelson Archives Center, National Museum of American History, Smithsonian Institution

Notes

Introduction

1 Logan Clendening, "The Doctor's Corner," *Endicott Daily Bulletin* (5 August 1943), 12.
2 Eli Clare, *Brilliant Imperfection: Grappling with Cure* (Durham: Duke University Press, 2017), 87.
3 David Wright, *Deafness* (New York: Stein and Day, 1969), 5.
4 Brenda Jo Brueggemann, *Lend Me Your Ear: Rhetorical Constructions of Deafness* (Washington, D.C.: Gallaudet University Press, 1999); Kristen C. Harmon, "Growing Up to Become Hearing: Dreams of Passing in Oral Deaf Education," in Jeffrey A. Brune and Daniel J. Wilson (eds.), *Disability and Passing: Blurring the Lines of Identity* (Philadelphia: Temple University Press, 2013), 167–198. For more on personal narratives and disability, see G. Thomas Couser, *Recovering Bodies: Illness, Disability, and Life Writing* (Madison: University of Wisconsin Press, 1997).
5 Christopher Krentz, *Writing Deafness: The Hearing Line in Nineteenth-Century American Literature* (Chapel Hill: University of North Carolina Press, 2007), 2–3.
6 Brenda Jo Brueggemann, "Articulating Betweenity: Literacy, Language, Identity, and Technology in the Deaf/Hard-of-Hearing Collection," in (eds.) H. Lewis Ulman, Scott Lloyd DeWitt, and Cynthia L. Selfe (Logan, UT: Computers and Composition Digital Press, 2013); Brenda Jo Brueggemann, *Deaf Subjects: Between Identities and Places* (New York: New York University Press, 2009).
7 Carol Padden and Tom Humphries, *Inside Deaf Culture* (Cambridge: Harvard University Press, 2009), 6.
8 Hillel Schwartz, *Making Noise: From Babel to the Big Bang and Beyond* (Cambridge, MA: MIT Press, 2011), 173. Citizenship, of course, is not static, and obligations of citizenship differ according to race, gender, and presumably disability. See Roger Smith, *Civic Ideals: Confronting Visions of Citizenship in U.S. History* (New Haven: Yale University Press, 1997); Michael Schudson, *The Good Citizen: A History of American Civic Life* (New York: Simon & Schuster, 1998).
9 While there is plenty of literature exploring the history of the normal and the norm, a thoroughly comprehensive analysis is Peter Cryle and Elizabeth

Stephens, *Normality: A Critical Genealogy* (Chicago: University of Chicago Press, 2017). See also Anna G. Creadick, *Perfectly Average: The Pursuit of Normality in Postwar America* (Amherst: University of Massachusetts Press, 2010); Caroline Warman, "From Pre-Normal to Abnormal: The Emergence of a Concept in Late Eighteenth-Century France," *Psychology and Sexuality* 1.3 (2010): 200–213; Elizabeth Stephens, "Normal," *Transgender Studies Quarterly* 1 (2014): 141–145; Julian B. Carter, *The Heart of Whiteness: Normal Sexuality and Race in America, 1880–1940* (Durham: Duke University Press, 2007). There are plenty of critiques of normalcy and normativity, especially in queer studies and critical disability studies: Lennard J. Davis, *Enforcing Normalcy* (London: Verso Books, 1995); Tanya Titchkosky and Rod Michalko (eds.), *Rethinking Normalcy: A Disability Studies Reader* (Toronto: Canadian Scholars' Press Inc., 2009); J. Jack Halberstam, *In Queer Time and Place: Transgender Bodies, Subcultural Lives* (New York: New York University Press, 2005); Rosemarie Garland-Thomson (ed.), *Freakery: Cultural Spectacles of the Extraordinary Body* (New York: New York University Press, 1996).

10 Cryle and Stephens, *Normality*, 19. See also Sarah E. Igo, *The Averaged American: Surveys, Citizens, and the Making of a Mass Public* (Cambridge, MA: Harvard University Press, 2008).

11 Creadick, *Perfectly Average*, 2.

12 Courtney Q. Shah, *Sex Ed, Segregated: The Quest for Sexual Knowledge in Progressive-Era America* (Rochester: University of Rochester Press, 2015), 133.

13 Cryle and Stephens, *Normality*, 319. See also Sarah F. Rose, *No Right to Be Idle: The Invention of Disability, 1840s–1930s* (Chapel Hill: University of North Carolina Press, 2017); Robert M. Buchanan, *Illusions of Equality: Deaf Americans in School and Factory, 1850–1950* (Washington, D.C.: Gallaudet University Press, 1999).

14 Douglas Baynton, *Forbidden Signs: American Culture and the Campaign Against Sign Language* (Chicago: University of Chicago Press, 1996); Jan Branson and Don Miller, *Damned for Their Difference: The Cultural Construction of Deaf People as Disabled* (Washington, D.C.: Gallaudet University Press, 2002).

15 Baynton, *Forbidden Signs*.

16 Branson and Miller, *Damned for their Difference*, 170. It's also worth pointing out that cultural constructions of "normal" can also have detrimental consequences, especially once given a privileged position. See Jonathan Sholl, "Nobody Is Normal," *Aeon* (31 January 2017), https://aeon.co/essays/is-it-time-to-abandon-the -medical-construct-of-being-normal/.

17 Esme Cleall, " 'Deaf to the Word': Gender, Deafness, and Protestantism in Nineteenth-Century Britain," *Gender and History* 25.3 (November 2013): 593.

18 Noga Arikha, "Deafness, Ideas, and the Language of Thought in the Late 1600s," *British Journal for the History of Philosophy* 13.2 (2005): 233–262.

19 The capital-*D* "Deaf" was coined in 1972 by the sociologist James Woodward to distinguish between deafness as an audiological affliction, and the community of individuals with hearing loss with a distinct culture grounded in the use of sign language. James Woodward, "Implications for Sociolinguistics Research among the Deaf," *Sign Language Studies* 1 (1972): 1–7.

20 Baynton, *Forbidden Signs*, 132.

21 Douglas C. Baynton, *Defectives in the Land: Disability and Immigration in the Age of Eugenics* (Chicago: University of Chicago Press, 2016); Janet Golden and

John T. Duffy, "'Normal Enough': Paula Patton, Intellectually Disabled Immigrant Children, and the 1924 Immigration Act," *Journal of Social History* shy098 (2019): 1–25.

22 For an overview of how proponents of methodological sign language signalled a preference for hearing culture, see R. A. R. Edwards, *Words Made Flesh: Nineteenth-Century Deaf Education and the Growth of Deaf Schools* (New York: New York University Press, 2012).

23 From the 1860s, "oralists" battled with "signers" over speech, deafness, and humanity, a battle that heightened in 1880 Second International Congress on the Education of the Deaf in Milan, which banned sign language use in the classroom. Oralism was taught exclusively until the late 1960s, when "total communication" (which claimed to offer the best pedagogical method for a deaf child) was introduced by Roy Kay Holcomb; the Deaf President Now movement of the 1980s eventually returned sign language for classroom instruction. On the movement, see John B. Christiansen and Sharon N. Barnartt, *Deaf President Now! The 1988 Revolution at Gallaudet University* (Washington. D.C.: Gallaudet University Press, 1995); Jack R. Gannon, *The Week the World Heard Gallaudet* (Washington, D.C.: Gallaudet University Press, 1989).

24 Davis, *Enforcing Normalcy*, 78.

25 On the history of oralism and how different professions defined deafness as a pathology, a deviance, a disability, and a sociocultural trait, see Marion Schmidt, "Genetic Normalcy and the Normalcy of Difference: Hereditary Deafness Research Throughout the 20th Century," Ph.D. dissertation, Johns Hopkins University, 2016. See also Esme Cleall, "Orientalising Deafness: Disability and Race in Imperial Britain," *Social Identities* 21.1 (Spring 2015): 22–36.

26 On medicalization and biomedicalization, see Peter Conrad, *The Medicalization of Society: On the Transformation of Human Conditions into Treatable Disorders* (Baltimore: Johns Hopkins University Press, 2007), and Adele E. Clarke, Laura Mamo, Jennifer Ruth Fosket, Jennifer R. Fishman, and Janet K. Shim (eds.), *Biomedicalization: Technoscience, Health, and Illness in the U.S.* (Durham: Duke University Press, 2010).

27 Susan Burch, *Signs of Resistance: American Deaf Cultural History, 1900 to World War II* (New York: New York University Press, 2002), 13.

28 Baynton, *Forbidden Signs*, 8.

29 Baynton, *Forbidden Signs*, 142.

30 John D. Wright to Irving S. Fusfield, 29 January 1924, Irving Fusfield Papers, Mss 30, Gallaudet University Archives, box 13, folder 36.

31 Hilda Tillinghast to Irving S. Fusfield, 3 February 1930, Irving Fusfield Papers, Mss 30, Gallaudet University Archives, box 16, folder 34. Tillinghast married Boyce R. Williams, a deaf man responsible for vocational rehabilitation programs for deaf Americans. For more on the Tillinghast family of educators, see John Vickery Van Cleve and Barry A. Crouch, *A Place of Their Own: Creating the Deaf Community in America* (Washington, D.C.: Gallaudet University Press, 1989).

32 Mara Mills, "Deafening: Noise and Engineering of Communication in the Telephone System," *Grey Room* 43 (2011): 118–143; Jaipreet Virdi, "Prevention and Conservation: Historicizing the Stigma of Hearing Loss, 1910-1940," *Journal of Law, Medicine, and Ethics* 45 (2017): 531–544.

33 "Hope: How Medical Help Also Moves Forward," *New York League of the Hard*

of Hearing Bulletin 12, no. 8–9, silver anniversary souvenir number (December 1935–January 1936).

34 Creadick, *Perfectly Average*, 60.

35 Donna McDonald, *The Art of Being Deaf: A Memoir* (Washington, D.C.: Gallaudet University Press, 2014), 39.

36 Daniel J. Wilson, "Passing in the Shadow of FDR: Polio Survivors, Passing, and the Negotiation of Disability," in Jeffrey A. Brune and Daniel J. Wilson (eds.), *Disability and Passing: Blurring the Lines of Identity* (Philadelphia: Temple University Press, 2013), 14.

37 Jeffrey A. Brune and Daniel J. Wilson, "Introduction," in Jeffrey A. Brune and Daniel J. Wilson (eds.), *Disability and Passing: Blurring the Lines of Identity* (Philadelphia: Temple University Press, 2013), 2.

38 Marie Hays Heiner, *Hearing Is Believing* (Cleveland: World Publishing Company, 1949), 29.

39 Heiner, *Hearing Is Believing*, 14.

40 Erving Goffman, *Stigma: Notes on the Management of Spoiled Identity* (New York: Simon & Schuster, 1963), 9.

41 Schachtel's address was delivered at the 36th Annual Conference of the Association of Better Business Bureau and published as *Current Problems in the Hearing Aid Industry* (Washington, D.C: Better Business Bureau, 1950).

42 Graeme Gooday and Karen Sayer, "Purchase, Use, and Adaptation: Interpreting 'Patented' Aids to the Deaf in Victorian Britain," in Claire L. Jones (ed.), *Rethinking Modern Prostheses in Anglo-American Commodity Cultures, 1820–1939* (Manchester: Manchester University Press, 2017).

43 "Australia's Ad Standard Board Rules Ad Saying 'Hearing Aids Can Be UGLY' Is Discriminatory," *Northern Virginia Resource Center*, 23 June 2015, http://www.nvrc.org/2015/06/australias-ad-standards-board-rules-ad-saying-hearing-aids-can-be-ugly-is-discriminatory/.

44 "Caitlyn Jenner Heading Towards Another Transition," *Terez Owens* (20 November 2017), https://terezowens.com/caitlyn-jenner-heading-towards-another-transition/.

45 Arthur J. Cramp to Bessie A. Chamberlin, 9 November 1931, Deafness Cures 1905–1973, box 181, folder 13, HHF-AMA.

46 Charles J. Gotthart to Arthur J. Cramp, 4 November 1925. Deafness Cures 1905–1973, box 181, folder 14, HHF-AMA.

47 "Wanted! 25,000 Sufferers from Deafness," *Columbus Dispatch*, 23 October 1927. Advertisement clipping, Deafness Cures 1905–1973, box 181, folder 13, HHF-AMA.

48 Miracle-Ear commercial, 1991, YouTube, https://www.youtube.com/watch?v=KpgfCsEy6r8/.

49 Miracle-Ear commercial, 1987, YouTube. https://www.youtube.com/watch?v=4IJwJ9Xlj-c/.

50 "How Deafness Can Be Cured," *New York Times* (1 January 1886), 6.

51 Francis L. Rogers, "Can Education End Deafness?" *Federation News* 3.8 (January 1935): 1.

52 National Institute on Deafness and Other Communication Disorders, National Institutes of Health, U.S. Department of Health & Human Services (15 December 2016), https://www.nidcd.nih.gov/health/statistics/quick-statistics-hearing/.

53 Most American aural surgeons were trained in Vienna, Berlin, Paris, or London, as part of their postgraduate education, especially by Adam Politzer (1835-1920) in Vienna, such as Clarence Blake, the first professor of otology at Harvard and chief of the aural department at Massachusetts Eye and Ear Infirmary. An 1879 letter from Blake to Politzer acknowledges the latter's influence: "We have every reason to be encouraged as to the standing of otology in America in the future and the cordial good feeling which exists among aurists in this country will do much to advance our branch of science. The aurists here seem always ready to acknowledge each others' good work and to help each other in study and in experiment." For more on the developments of American aural surgery, see Neil Weir, *Otolaryngology: An Illustrated History* (London: Butterworths, 1990), 184.

54 Hannah Joyner, *From Pity to Pride: Growing Up Deaf in the Old South* (Washington, D.C.: Gallaudet University Press, 2004).

55 Joyner, *From Pity to Pride*, 7.

56 A. B. Surditas, "The Uncertainties of Aural Surgery," *Association Medical Journal* 1.20 (20 May 1853): 447-448.

57 Eric Boyle, *Quack Medicine: A History of Combating Health Fraud in Twentieth-Century America* (Santa Barbara: Prager, 2013), xxii.

58 Hallowell Davis (ed.), *Hearing and Deafness: A Guide for Laymen* (New York: Reinhart & Company, Inc., 1954), 165.

59 Arthur J. Cramp, "Deafness-Cure Quackery and Pseudo-Medicine," *Volta Review* 28.9 (September 1926): 497.

60 Peter N. Stearns, *Fat History: Bodies and Beauty in the Modern West* (New York: New York University Press, 2002); Andrew Abbott, *The System of Professions: An Essay on the Division of Expert Labor* (Chicago: University of Chicago Press, 1988).

61 Clara B. Seaman to AMA, 6 October 1919, Deafness Cures 1905-1973, box 181, folder 14, HHF-AMA.

62 Max A. Goldstein, *Problems of the Deaf* (St. Louis: Laryngoscope Press, 1933), 461.

63 Mara Mills, "Deafness," in David Novak and Matt Sakakeeny (eds.), *Keywords in Sound* (Durham: Duke University Press, 2015).

64 Laura Mauldin, *Made to Hear: Cochlear Implants and Raising Deaf Children* (Minneapolis: University of Minnesota Press, 2016).

65 Harlan Lane's work is a classical exposition of the Deaf culture vs. deaf medicalization issue. See Lane, *The Mask of Benevolence: Disabling the Deaf Community* (New York: Vintage Books, 1993).

66 See Burch, *Signs of Resistance*; Nielsen, *A Disability History of the United States* (Boston: Beacon Press, 2012).

67 Melissa Malzkuhn, "Compromising for Agency: The Role of the NAD During the American Eugenics Movement, 1880-1940," in Brian H. Greenwald and Joseph J. Murray (eds.), *In Our Own Hands: Essays in Deaf History 1780-1970* (Washington, D.C.: Gallaudet University Press, 2016), 171-92; Brian H. Greenwald, "Alexander Graham Bell Through the Lens of Eugenics 1883-1922," Ph.D. dissertation, George Washington University, 2006; Brian H. Greenwald, "The Real 'Toll' of A. G. Bell: Lessons About Eugenics," *Sign Language Studies* 9.3 (2009): 258-265.

68 Virdi, "Prevention and Conservation," 532.

69 Maren Klawiter, *The Biopolitics of Breast Cancer: Changing Cultures of Disease and Activism* (Minneapolis: University of Minnesota Press, 2008), 59.

70 Arthur J. Cramp, *Nostrums and Quackery*, vol. 2 (Chicago: Press of American Medical Association, 1921), 109.

71 Boyle, *Quack Medicine*, 65–69.

72 Mara Mills, "Another Etymology for 'Bionic': Hearing Aids and Disability History at Kent State," *Rittenhouse: Journal of the American Scientific Instrument Enterprise* 22 (2008): 47–69.

73 Letter from E. F. McDonald Jr. to Howard A. Carter, 24 January 1952, CID Series 3, box 17.

74 Roland Marchand, *Advertising the American Dream: Making Way for Modernity, 1920–1940* (Berkeley: University of California Press, 1985).

75 Nancy Tomes, *Remaking the American Patient: How Madison Avenue and Modern Medicine Turned Patients into Consumers* (Chapel Hill: University of North Carolina Press, 2016); Nancy Tomes, "Merchants of Health: Medicine and Consumer Culture in the United States, 1900–1940," *Journal of American History* 88.2 (September 2001): 519–547; William Leach, *Land of Desire: Merchants, Power, and the Rise of a New American Culture* (New York: Vintage Books, 1993).

76 The Envoy Esteem Implant is engineered for moderate to severe hearing loss and is implanted in the skin behind the ear. It is more invisible than conventional hearing aids. "29 Years Old and Hearing Myself for the 1st time!" YouTube (uploaded by Sloan Churman, 29 September 2011), https://www.youtube.com/watch?v=LsOo3jzkhYA/.

77 Sarah Churman, *Powered On: On the SOUNDS I Choose to Hear and the NOISE I Don't* (Pensacola: Indigo River Publishing, 2012).

78 For an example of commentary on this narrative, see Lilit Marcus, "Why You Shouldn't Share Those Emotional 'Deaf Person Hears for the First Time' Videos," *Wire*, 28 March 2014, http://www.thewire.com/politics/2014/03/why-you-shouldnt-share-those-emotional-deaf-person-hears-forthe-first-time-videos/359850/.

79 Stuart Blume, *The Artificial Ear: Cochlear Implants and the Culture of Deafness* (New Brunswick, NJ: Rutgers University Press, 2010); Paddy Ladd, *Understanding Deaf Culture: In Search of Deafhood* (Clevedon: Multilingual Matters, 2003).

80 Lydia Denworth, "Science Gave My Son the Gift of Sound," *Time* (25 April 2014), http://time.com/76154/deaf-culture-cochlear-implants/.

81 Jennifer Esmail, *Reading Victorian Deafness: Signs and Sounds in Victorian Literature and Culture* (Athens: Ohio University Press, 2013).

82 Clare, *Brilliant Imperfection*, 10, 25–26.

Chapter 1

1 Quoted in Elliot Forbes, *Thayer's Life of Beethoven, Volume 1*, rev. ed. (Princeton: Princeton University Press, 1967), 283. "Frank" refers to Peter Frank, the director of the general hospital in Vienna.

2 Pierpont F. Bowker, *The Indian Vegetable Family Instruction* (Utica: Jared Doolittle, 1851), reprinted in Adelaide Hechtlinger, *The Great Patent Medicine Era, or, Without Benefit of Doctor* (New York: Grosset & Dunlap Inc., 1970), 194.

3 *The Pretensions of Dr. Alexander Turnbull, to Cure Deaf-Dumbness and All Diseases of the Eye and Ear* (Charleston: Walker and James Press, 1854), American Antiquarian Society Archives, Misc. Pams. 1854 Pret.

4 Sarah Helen DeKroyft's (1818–1915) letter of gratitude to Mrs. Fillmore is dated 11 August 1852 and was reprinted in *The Cleveland Herald*, the *New Orleans Picayune*, and New Orleans' *Daily Delta*, among other periodicals. See "Interesting Letter to Mrs. Fillmore," *The Cleveland Herald* (12 January 1853), 2. De Kroyft also outlined her emotions dealing with blindness in *A Place in Thy Memory* (New York: John F. Trow, 1851).

5 Joseph Henry Johnson to Emma Johnson, 30 January 1854, Johnson Family Papers, 1798–1943, Rubenstein Rare Books and Manuscripts Library, Duke University, box 1, c.1. On the history of the Johnson family and the Talladega (later, Alabama) Institute for the Deaf, Dumb, and Blind, see Thomas McAdory Owen, *History of Alabama and Dictionary of Alabama Biography*, vol. 3 (Chicago: S. J. Clarke Publishing Company, 1921), 912.

6 Joseph Henry Johnson to Emma Johnson, 30 January 1854, Johnson Family Papers, 1798–1943, Rubenstein Rare Books and Manuscripts Library, Duke University, box 1, c.1.

7 Robert Hill Couch and Jack Hawkins Jr., *Out of Silence and Darkness: The History of the Alabama Institute for Deaf and Blind 1858–1953* (Troy, AL: Troy State University Press, 1983), 21.

8 Ira Rutkow, *Seeking the Cure: A History of Medicine in America* (New York: Scribner, 2010), 45, 32.

9 Seth Stein LeJacq, "The Bounds of Domestic Healing: Medical Recipes, Storytelling, and Surgery in Early Modern England," *Social History of Medicine* 26.3 (2013): 453; Susan E. Cayleff, *Nature's Path: A History of Naturopathic Healing in America* (Baltimore: Johns Hopkins University Press, 2016), 17.

10 Jeremy Agnew, *Medicine in the Old West: A History, 1850–1900* (Jefferson, NC: McFarland & Company, 2010), 127.

11 John C. Burnham, *Health Care in America: A History* (Baltimore: Johns Hopkins University Press, 2015), 45.

12 Hatfield, *Encyclopedia of Folk Medicine*, 111–112.

13 Herbert C. Covey, *African-American Slave Medicine: Herbal and Non-Herbal Treatments* (Plymouth: Lexington Books, 2007), 97; Patricia D. Schafer, "A Manual of Cherokee Herbal Remedies: History, Information, Identification, Medicinal Healing," master's thesis, Indiana State University, 1993, 68.

14 Lady Jane Francesca Wilde, *Ancient Cures, Charms, and Usages of Irelands: Contributions to Irish Lore* (London: Ward and Downey, 1890), 13. Lady Wilde's publication heavily relied on William Wilde's manuscript, "History of Irish Medicine and Popular Cures" [1840s], which is housed at the University College Dublin Joyce Library Special Collections. See also Gabrielle Hatfield, *Encyclopedia of Folk Medicine: Old World and New World Traditions* (Santa Barbara: ABC-CLIO, 2004), 112.

15 LeJacq, "The Bounds of Domestic Healing," 456.

16 Burnham, *Health Care in America*, 31; John C. Burnham, *How Superstition Won and Science Lost: Popularizing Science and Health in the United States* (New Brunswick, NJ: Rutgers University Press, 1987).

17 For more on Punjabi folklore and the evil eye, see Azher Hameed Qamar, "Belief in the Evil Eye and Early Childcare in Rural Punjab, Pakistan," *Asian Ethnology* 75.2 (2016): 397–418.

18 Burnham, *Health Care in America*, 31.

19 Edward Topsell, *The History of Four-Footed Beasts and Serpents* (London: Printed by E. Cotes, for G. Sawbridge [etc.], 1658), 65.

20 Topsell, *The History of Four-Footed Beasts and Serpents*, 65.

21 Nancy Catford, "If One Be Deafe or Thicke of Hearing: Ancient Cures for Deafness," *Silent World* 3.7 (December 1948): 208–210.

22 Hugh Smythson, M.D., *The Compleat Family Physician; or Universal Medical Repository* (London: Harrison and Co.,1781).

23 Thomas F. Branston, *The Druggist's Hand-Book of Practical Receipts: A Manual for the Use of the Chemist and Medical Practitioner* (Liverpool: Edward Howell, Publisher, 1853).

24 LeJacq, "The Bounds of Domestic Healing," 454; Elaine Leong, "Making Medicines in the Early Modern Household," *Bulletin of the History of Medicine* 82 (2008): 145–168.

25 On women and the activity of compiling recipe books, see Mary Fissell, "Introduction: Women, Health, and Healing in Early Modern Europe," *Bulletin of the History of Medicine* 82 (2008): 1–17; Montserrat Cabré, "'Women or Healers?' Household Practices and the Categories of Health Care in Late Medieval Iberia," *Bulletin of the History of Medicine* 82 (2008): 18–52.

26 Beinecke Osborn c663, Beinecke Rare Book and Manuscript Library, quoted in Eve Houghton, "Recipes in Manuscript Miscellanies" (20 October 2016), *The Recipes Project*, https://recipes.hypotheses.org/8572/.

27 Recipe and home remedy book, c. 1896, Manuscripts, Rubenstein Rare Books and Manuscripts Library, Duke University, sec A, box 110.

28 Elizabeth Hammond, *Modern Domestic Cookery and Useful Receipt Book*, 3rd ed. (London: Dean & Munday, Threadneedle-Street, 1819), 224.

29 Mrs. A. P. Hill, *Mrs. Hill's New Cook Book* (New York: Carleton, 1872 [1867]), 391, reprinted in Annabella P. Hill, *Mrs. Hill's Southern Practical Cookery and Receipt Book* (Columbia: University of South Carolina Press, 1995).

30 Andreas Markides, "Some Unusual Cures for Deafness," *Journal of Laryngology and Otology* 96.6 (June 1982): 487.

31 *Family Receipts: Being a Compilation from Several Publications* (Syracuse: L. H. Redfield's Bookstore, 1840), 59.

32 William Wright, *The Aurist* 1 (31 March 1825), 3.

33 Alexander Turnbull, *On the Medical Properties of the Natural Order Ranunculaceæ* (London: Longman, Rees, Orme, Browne, Green & Longman, 1835).

34 William Banting, *Letter on Corpulence, Addressed to the Public* (London: Harrison, 1863), 12.

35 Banting, *Letter on Corpulence*, 16.

36 The series was instrumental in the passage of the Pure Food and Drug Act of 1906, specifying that a patent drug's label should not be false or misleading. However, as a 1911 Supreme Court ruling clarified, the act was limited to identifying ingredients used in proprietary medicines, not to its claims of efficacy. Similar crusades against fraudulent assertions of patent medicine vendors were investigated and reported in the *Ladies' Home Journal*, by state health officials, local

newspapers, the Better Business Bureau, and the American Medical Association's Propaganda Department.

37 Richard J. Dunlap to *JAMA*, 16 November 1926, Department of Investigation Records, box 155, folder 5, HHF-AMA.

38 W. O. Coffee pamphlet, n.d., Department of Investigation Records, box 155, folder 5, HHF-AMA.

39 Advertisement for W. O. Coffee, 1928, Department of Investigation Records, box 155, folder 5, HHF-AMA.

40 Despite Selfridge's training, it is not clear how much homeopathy influenced, or was used in, his clinical practice. He was later employed at the Southern Pacific Hospital in San Francisco, where he specialized in otolaryngology and was supposedly the first surgeon in San Francisco to perform a tonsillectomy. Jonathan Davidson, *A Century of Homeopaths: Their Influence on Medicine and Health* (New York: Springer Books, 2014), 10.

41 Grant Selfridge, "Present Status of Vitamins in Relation to Eighth Nerve and Conduction Deafness," *Archives of Otolaryngology* 34.1 (1941): 125–140.

42 Grant Selfridge, "Eighth Nerve and Conduction Deafness," *California and Western Medical Journal* 52.5 (May 1940): 214–217.

43 G. Grant, "On the Physical Exploration of the Ear by Means of the Auriscope," *Medical and Surgical Reporter* 10 (1857): 449.

44 W. F. Bynum, *Science and Practice of Medicine in the Nineteenth Century* (Cambridge: Cambridge University Press, 1994), 114.

45 Leslie J. Reagan, *Dangerous Pregnancies: Mothers, Disabilities, and Abortion in Modern America* (Berkeley: University of California Press, 2010).

46 D. B. St. John Roosa, *A Practical Treatise on the Diseases of the Ear, Including a Sketch of Aural Anatomy and Physiology*, 6th ed. (New York: William Wood & Company, 1885), 590.

47 Mara Mills, "Deafening: Noise and the Engineering of Communication in the Telephone System," *Grey Room* 42 (2011): 118–143; Jaipreet Virdi and Coreen McGuire, "Phyllis M. Tookey Kerridge and the Science of Audiometric Standardization in Britain," *British Journal for the History of Science* 51.1 (2018): 127.

48 Hannah Joyner, *From Pity to Pride: Growing up Deaf in the Old South* (Washington, D.C.: Gallaudet University Press, 2004), 14.

49 Harald Lindenov, *The Etiology of Deaf-Mutism with Special Reference to Heredity* (Copenhagen: Einar Munksgaard, 1945), 10–11.

50 Thomas J. Pettigrew, *On Superstitions Connected with the History and Practice of Medicine and Surgery* (Philadelphia: E. Barrington and G. D. Haswell, 1844), 131.

51 Jennifer Esmail, *Reading Victorian Deafness: Signs and Sounds in Victorian Literature and Culture* (Athens: Ohio University Press, 2013), 166.

52 James Keene, *The Causes and Treatment of Deafness; Being a Manual of Aural Surgery for the Use of Students and Practitioners of Medicine* (London: Robert Hardwicke, 1873), 2.

53 John Harrison Curtis, *An Essay on the Deaf and Dumb; Shewing the Necessity of Medical Treatment in Early Infancy, with Observation on Congenital Deafness* (London: Longman, Rees, Orme, Brown and Green, 1829).

54 Mary Wilson Carpenter, *Health, Medicine, and Society in Victorian England* (Santa Barbara: ABC-CLIO, 2010), 115.

55 While there are several strains that can cause bacterial meningitis, they usually

fall into four categories: (1) *Streptococcus pneumoniae* (pneumococcus), (2) *Neisseria meningitidis* (meningococcus), (3) *Haemophilus influenzae* (haemophilus), and (4) *Listeria monocytogenes* (listeria).

56   Kenneth L. Tyler, "A History of Bacterial Meningitis," in *Handbook of Clinical Neurology*, vol. 95 (3rd series), *History of Neurology*, ed. S. Finger, F. Boller, and K. L. Tyler (Edinburgh: Elsevier, 2010), 417.

57   Thomas Willis, *London Practice of Physick or the Whole Practical Part of Physick contained in the works of Dr. Willis* (London: Baffet and Crook, 1685).

58   Jack R. Gannon, *Deaf Heritage: A Narrative History of Deaf America*, rev. ed. (Washington, D.C.: Gallaudet University Press, 2012), 2.

59   Astley Cooper, "Observations on the Effects Which Take Place from the Destruction of the Membrana Tympani of the Ear," *Philosophical Transactions of the Royal Society of London* 90 (1800): 1–21; Astley Cooper, "Farther Observations on the Effects Which Take Place from the Destruction of the Membrana Tympani of the Ear," *Philosophical Transactions of the Royal Society of London* 91 (1801): 435–450.

60   J. Rimmer, C. E. B. Giddings, and Neil Weir, "History of Myringotomy and Grommets," *Journal of Laryngology and Otology* 121 (2007): 912.

61   *The Mirror of Literature, Amusement, and Instruction*, 29 (17 May 1823), 455.

62   Keene, *The Causes and Treatment of Deafness*, 4.

63   *Report of the Trial of Thomas Bent Hodgson, Esq., and others charged with a conspiracy: at the court of King's Bench, Guildhall, London, on Wednesday & Thursday the 21st and 22nd of December 1831 by a special jury before the Right Hon. Lord Tenterden* (Printed by James & Luke G. Hansard & Sons, 1831), 452. The case followed an indictment against Thomas R. Hodgson, William Young, Ann Young, and Nancy Watson, charging them with having conspired to manipulate William Stephenson's will. The defendants argued that Stephenson's fortune was duly executed, and that Turnbull, as Stephenson's physician, manipulated Stephenson to rewrite his will. The verdict found the defendants not guilty of conspiracy.

64   Bartholomeus Eustachius, *Epistola de auditus organius* (Examination of the organ of hearing) (1562). This treatise is possibly the first on the anatomy of the ear.

65   Steven H. Hale, "Antonio Maria Valsalva (1666–1723)," *Clinical Medicine & Research* 3.1 (2005): 35–38.

66   Anne T. Quartararo, *Deaf Identity and Social Images in Nineteenth-Century France* (Washington, D.C.: Gallaudet University Press, 2008), 22.

67   Jean-Marc Gaspard Itard, *Traité des Maladies de l'Oreille et de l'Audition* (Paris: Méquignon-Marvis, 1821).

68   Roger Shattuck, *The Forbidden Experiment: The Story of the Wild Boy of Aveyron* (New York: Kodansha International, 1980).

69   George Day, "On the Late Efforts in France and Other Parts of Europe to Restore the Deaf and Dumb to Hearing," *American Journal of Sciences and Arts* 30 (1836): 316.

70   On the practice of coroner's inquests, see Ian A. Burney, *Bodies of Evidence: Medicine and the Politics of the English Inquest, 1830–1926* (Baltimore: Johns Hopkins University Press, 2000).

71   "Inquests in Middlesex: Evidence of Witnesses at the Inquest on the body of Joseph Hall," *Lancet* 32 (3 August 1839): 690.

72   Alfred Barker, *A Treatise on Deafness and Rupture, with the New Mode of Cure*

(London: published by the author and sold by Gilbert, 49 Paternoster Row, [1855]), 6.

73 David Johnson to Edward Johnson, 25 February 1854, David Johnson Papers, 1810–1985, South Caroliniana Library, University of South Carolina, Manuscripts P, folder 5.

74 E. S. Aborn, *Practical Facts by a Practical Man, or the Pathway to Health and Happiness* (Philadelphia: Inquirer Book and Job Printing Office, 1867).

75 "Varieties," *The Titusville Herald*, 16 February 1867, 3.

76 *Mexico Independent and Deaf Mute's Journal*, 10.4 (28 November 1872): 1.

77 "Hypnotism as a Cure for Deafness," *American Annals for the Deaf* 40.2 (April 1895): 163.

78 "Deafness Cures (?)," *Rochester Advocate* 42 (15 March 1922), 6.

Chapter 2

1 "Aids for Deaf Persons," *Michigan Farmer* (13 June 1882): 13, 24.

2 B. Alex Randall, "Some Facts About Artificial Drumheads and Other Aids of Hearing," *Monthly Cyclopaedia of Practical Medicine* 8 (1905): 301.

3 Harry Stephen Keeler, *The Man with the Magic Eardrums* (Vancleave, MS: Ramble House, 2010; originally published New York: Dutton, 1939), 27.

4 James Harvey Young, *The Medical Messiahs: A Social History of Health Quackery in Twentieth-Century America*, expanded ed. (Princeton: Princeton University Press, 1992), 13.

5 Claire L. Jones, "Introduction: Modern Prostheses in Anglo-American Commodity Cultures," in Claire L. Jones (ed.), *Rethinking Modern Prostheses in Anglo-American Commodity Cultures, 1820–1939* (Manchester: Manchester University Press, 2017), 13.

6 "Deafness," advertisement for F. C. Rein, printed in *London Standard* (13 November 1869).

7 Jones, "Introduction," 4.

8 David M. Turner, "Disability and Prosthetics in Eighteenth- and Early Nineteenth-Century England," in Mark Jackson (ed.), *Routledge History of Disease* (London: Routledge, 2017), 302.

9 Graeme Gooday and Karen Sayer, *Managing the Experience of Hearing Loss in Britain, 1830–1930* (London: Palgrave Macmillan, 2017), 65–66.

10 Gooday and Sayer, *Managing the Experience of Hearing Loss*, 4.

11 Globe Optical Company, *Special Catalogue of Hearing Instruments Imported & Manufactured by Globe Optical Company of Boston* (Boston: c. 1890), Trade catalog collection, Hagley Library and Archives.

12 Jaipreet Virdi, "Between Cure and Prosthesis: 'Good Fit' in Artificial Eardrums," in Claire L. Jones (ed.), *Rethinking Modern Prostheses in Anglo-American Commodity Cultures, 1820–1939* (Manchester: Manchester University Press, 2017), 51.

13 J. M. Churchill, "Mr. Toynbee on Deafness," *Medical Times and Gazette* (1853): 281. See also Rebecca J. Rosen, "Why Are Glasses Perceived Differently than Hearing Aids?" *Atlantic* (3 December 2013).

14 Willis (William V.) and Company, "Instruments to Assist the Hearing," 6th ed., (Philadelphia: c. 1900), trade catalog collection, Hagley Library and Archives.

15 Gooday and Sayer, *Managing the Experience of Hearing Loss*, 2.

16 Amelia Woods's conversation tube and pouch, Waterloo Region Museum/Doon Heritage Crossroads (970.066.001). For more on these pouches as examples of user adaptation, see Graeme Gooday and Karen Sayer, "Purchase, Use, and Adaptation: Interpreting 'Patented' Aids to the Deaf in Victorian Britain," in Claire L. Jones (ed.), *Rethinking Modern Prostheses in Anglo-American Commodity Cultures, 1820–1939* (Manchester: Manchester University Press, 2017), 41.

17 Turner, "Disability and Prosthetics," 315–316.

18 Dorothy Brett, "My Long and Beautiful Journey," Dorothy Brett unpublished memoirs, 16, DBC, box 1, folder 6.

19 Brett, "My Long and Beautiful Journey," 56.

20 Telegram from Edward VII to Esher, 15 October 1902. Dorothy Brett Collection 1898–1968, box 27, folder 1, Harry Ransom Center, The University of Texas at Austin.

21 Sean Hignett, *Brett: From Bloomsbury to New Mexico. A Biography* (London: Hodder and Stoughton, 1984), 24.

22 Quoted in Hignett, *Brett*, 50.

23 Quoted in Hignett, *Brett*, 60.

24 Dorothy Brett to Bertrand Russell, 26 August 1918, Bertrand Russell Archives, RA1, Box 5, folder 5 McMaster University.

25 Bertrand Russell to Dorothy Brett, 30 August 1918. Bertrand Russell, *The Autobiography of Bertrand Russell*, 3 vols. (London: Allen & Unwin, 1967–69), 92–93.

26 Brett, "My Long and Beautiful Journey," 114.

27 Katherine Mansfield to Dorothy Brett, 29 August 1920, in Katherine Mansfield, *The Collected Letters of Katherine Mansfield*, vol. 4, *1920–1921*, ed. Vincent O'Sullivan and Margaret Scott (Oxford: Clarendon Press, 1996), 21.

28 Brett, "My Long and Beautiful Journey," 120.

29 "Christmas Morning," 25 December 1967, DBC, box 1, folder 13.

30 Hignett, *Brett*, 53.

31 Dorothy Brett, *Lawrence and Brett: A Friendship* (Santa Fe, NM: Sunstone Press, 2006), 169.

32 Brett, *Lawrence and Brett*, 163.

33 Brett, *Lawrence and Brett*, 187.

34 DBC, box 1, folder 5, MS 11; Mable Dodge Luhan Papers, box 196 YCAL MSS 196, Yale Collection of American Literature, Beinecke Rare Book and Manuscript Library.

35 Typescript draft of the foreword to Brett's book, n.d. DBC, box 1, folder 13.

36 Engraving of James Hutton, mezzotint by J. R. Smith (1786) after original painting by R. Cosway, Science Museum/Victoria & Albert Museum.

37 Daniel Benham, *Memoirs of James Hutton; Comprising the Annals of His Life, and Connection with the United Brethren* (London: Hamilton, Adams, & Co., 1856), 590.

38 W. P. Zuber, Library of Congress Prints and Photographs Division, LOT 11533–14.

39 The last two images are sourced from GettyImages: "Family gathering outside, ca. 1903," photo by Kirn Vintage Stock, and "1890s 1900s Portrait Senior Couple Seated in Garden Man Holding Ear Trumpet," photographed by H. Armstrong Roberts.

40 The acoustic gain measurements were made in an anechoic chamber using a

Knowles' Electronics Manikin for Acoustic Research. See Cathy C. Sarli et al., "19th-Century Camouflaged Mechanical Hearing Devices," *Otology and Neurotology* 24.4 (July 2003): 691–698.

41 As noted on the Hearing Aid Museum, www.hearingaidmuseum.com/.

42 Edith Ella Baldwin, "A Talk of Paris Art Days," Edith Ella Baldwin Papers, 1848–1920, Rubenstein Rare Books and Manuscripts Library, Duke University, box 7.

43 Joseph Toynbee, *The Diseases of the Ear: Their Nature, Diagnosis, and Treatment* (London: J. M. Churchill, 1860), 172.

44 Virdi, "Between Cure and Prosthetic," 52.

45 Quoted in James Yearsley, "On a New Mode of Treating Deafness," *Lancet* 52.1 (July 1848): 10–11. See also James Yearsley, *On the Artificial Tympanum: A New Mode of Treating Deafness*, 10th ed. (London: John Churchill & Sons, 1869), 33–34. While simultaneous developments were recorded by German aural practitioners Carl Gustav Linke in 1845 and Dr. Erhard of Berlin in 1849, none of the men was aware of the others' creation, and each claimed to be the first to employ cotton wool as a method for improving hearing.

46 Jaipreet Virdi, "Priority, Piracy, and Printed Directions: James Yearsley's Patenting of the Artificial Tympanum," *Technology and Innovation: Proceedings of the National Academy of Inventors* 16.2 (2014): 145–54. Until 1939, Toynbee's design was still being manufactured and sold on request by the American Optical Company of Chicago, Gordon Berry to Dr. Barton, 20 July 1939, Hearing Aids, 1907–1974, box 345, folder 3, HFF-AMA.

47 Thomas Barr, *Manual of Diseases of the Ear*, 2nd ed. (Glasgow: James Maclehose and Sons, 1896), 88.

48 Virdi, "Between Cure and Prosthetic," 57.

49 Thomas Hawksley, *Catalogue of Otoacoustical Instruments and Aids to the Deaf*, 3rd ed. (London: John Bale, Sons & Danielson, Ltd., 1895), 75.

50 John Nottingham, *Diseases of the Ear: Illustrated by Clinical Observations* (London: John Churchill, 1857), 86.

51 Wilson became president after a committee of public-spirited citizens asked him to be a Democratic school board candidate in 1906 because "they said members of the School Board ought to come from the ranks of the best citizens, men seeking for no advertisement and having no ax to grind." "Democratic Nominees for School Trustees, Choice of People," *Courier-Journal*, 2 November 1908, 8.

52 "Paid $8,500 For Back Yard In St. James Court," *Courier-Journal*, 5 September 1903, 9.

53 Takahiro Ueyama, *Health in the Marketplace: Professionalism, Therapeutic Desires, and Medical Commodification in Late-Victorian London* (Palo Alto, CA: Society for the Promotion of Science and Scholarship, 2010), 15.

54 Ueyama, *Health in the Marketplace*, 70.

55 James Harvey Young, *The Toadstool Millionaires: A Social History of Patent Medicines in America Before Federal Regulation* (Princeton: Princeton University Press, 1961), 166.

56 Beginning in 1903, Dakhyl would be tied up in a series of libel litigation against Henry Labouchère, owner and editor of *Truth* magazine for that statement. (*Dakhyl v. Labouchere* (1908) 2 KB 325 n). See also "New Reading of the Libel Law," *The Guardian*, 24 June 1908, 6.

57 "The Cure for Deafness," advertisement in *Daily Colonist* (2 November 1901).

58 J. H. Nicholson, *Nicholson's Patented Artificial Eardrums Restore Hearing to the Deaf* (London, 1890), British Library.

59 "Hope for the Deaf," advertisement in *Macmillan's Magazine* 58 (1888): 512.

60 Anonymous, *Exposures of Quackery: Being a Series of Articles upon, and Analyses of, Numerous Patent Medicines*, vol. 2 (London: Savoy Press, 1897), 49–51.

61 "Hope for the Deaf," advertisement in *South Australian Register* (7 July 1890).

62 October 1901, trial of John Nicholson and Henry Thomas Richards, t19011021-728, *Old Bailey Proceedings Online*.

63 "The Deaf Hear" advertisement in *Illustrated London News* (12 March 1898): 388.

64 Erik Larson brilliantly narrates the saga of Crippen in his book, *Thunderstruck* (New York: Crown/Archetype, 2006).

65 Evan Yellon, *Surdus in Search of His Hearing: An Exposure of Deafness Quacks, Frauds on the Deaf, and a Reliable Guide to the Best Means of Help for the Deaf* (London: Evan Macleod, 1910); *Exposures of Quackery: Being a Series of Articles upon, and Analyses of, Numerous Patent Medicines*, vol. 2 (London: Savoy Press, 1897).

66 Susan Burch, *Signs of Resistance: American Deaf Cultural History, 1900 to World War II* (New York: New York University Press, 2002), 31.

67 Lori A. Loeb, *Consuming Angels: Advertising and Victorian Women* (Oxford: Oxford University Press, 1994), 53–54.

68 Roland Marchand, *Advertising the American Dream: Making Way for Modernity 1920–1940* (Berkeley: University of California Press, 1985), 11.

69 George A. Akerlof and Robert J. Shiller, *Phishing for Phools: The Economics of Manipulation and Deception* (Princeton, NJ: Princeton University Press, 2015), 48.

70 Jeffrey L. Cruikshank and Arthur W. Schulz, *The Man Who Sold America: The Amazing (but True!) Story of Albert D. Lasker and the Creation of the Advertising Industry* (Boston, MA: Harvard Business Press, 2010), 52.

71 Edd Applegate, *The Rise of Advertising in the United States: A History of Innovation to 1960* (Lanham, MD: Scarecrow Press, 2012), 87.

72 Thomas Richards, *The Commodity Culture of Victorian England: Advertising and Spectacle, 1851–1914* (Stanford: Stanford University Press, 1990), 196.

73 "Court Paragraphs," *The Courier-Journal*, 6 March 1903, 10.

74 Wilson Ear Drum advertisement, "Lost the Point," *Century Illustrated Magazine* 60.6 (October 1900): 36.

75 Loeb, *Consuming Angels*, 183.

76 *Official Gazette of the United States Patent Office*, vol. 309 (April 1923), 50.

77 "Don't Shout!" Morley Ear Phone pamphlet, 1900, Trade catalog collection, Hagley Library and Archives.

78 W. S. Found to Arthur J. Cramp, 27 September 1932, Deafness Cures 1905–1973, box 183, folder 13, HHF-AMA, 5-1973, box 183, folder 13, HHF-AMA.

79 Richards, *The Commodity Culture of Victorian England*, 183.

80 Eric Boyle, *Quack Medicine: A History of Combating Health Fraud in Twentieth-Century America* (Santa Barbara: Prager, 2013).

81 Yellon, *Surdus in Search of His Hearing*, 16.

82 A.O. Leonard form letter, n.d. "Leonard, A. O., Correspondence, 1915–1938," box 469, folder 16, HHF-AMA.

83 Robert Harry Scott to Arthur J. Cramp, 17 November 1915, Department of Investigation Records, box 469, folder 16, HHF-AMA.

84 On 10 March 1920, a suit was filed for the seizure and condemnation of Leonard Ear Oil. Leonard never appeared for his hearing, so all seized bottles were destroyed by court order. The U.S. Department of Agriculture issued four additional notices of judgment against the oil, especially after advertisements appeared in the Pacific Coast region in 1921. On 3 July 1936, the FTC issued another complaint against Leonard, this time for manufacturing the ear oil and selling it in interstate commerce. Further, the complaint alleged the company had a historical tendency to deceive buyers and divert trade unfairly from competitors, in violation of the FTC's rules of fair business practice. No records of the Leonard company surface after this period.

85 James S. Mills to AMA, 11 October 1915, Deafness Cures 1905–1973, box 183, folder 13, HHF-AMA.

86 W. L. Holwick to AMA, 16 January 1929, Deafness Cures 1905–1973, box 183, folder 13, HHF-AMA.

87 A. P. Hilton to AMA, 31 July 1919, Deafness Cures 1905–1973, box 181, folder 14, HHF-AMA.

88 Letter from Alvin P. Cameron to Dr. Keefe, 9 May 1919, HHF-AMA, box 183, Way Folder.

89 George W. Prohaska to AMA, 27 October 1927, Department of Investigation Records, box 469, folder 16, HHF-AMA.

90 Cramp, *Nostrms and Quackery*, vol. 2, 131.

91 E. E. Bullis to J. M. Wynn, 22 December 1937, Albert D. Lasker Papers, box 11, folder 18. Huntington Library.

92 Judge Thornton W. Sargent to Arthur J. Cramp, 8 November 1926, Deafness Cures 1905–1973, box 183, folder 13, HHF-AMA.

93 Leonard F. Weiss to Arthur J. Cramp, 21 July 1921, Deafness Cures 1905–1973, box 181, folder 14, HHF-AMA.

94 Elise Wicken to Arthur J. Cramp, 17 March 1930, Deafness Cures 1905–1973, box 183, folder 13, HHF-AMA.

95 John W. Palmer to AMA, 9 January 1928, Deafness Cures 1905–1973, box 183, folder 13, HHF-AMA.

96 T. A. Casey to Arthur J. Cramp, 18 September 1934, Deafness Cures 1905–1973, box 183, folder 13, HHF-AMA.

97 "Deafness Cure-Fakes," *American Annals of the Deaf* 61 (1916): 195.

98 John Dutton Wright, *The Little Deaf Child: A Book for Parents* (New York: Wright Oral School, 1929), 155.

99 Different variations of artificial eardrums nevertheless emerged throughout the century. In 1950, *Life* magazine published a glossy article heralding a new opportunity for deaf readers—a "Plastic Eardrum." "Artificial eardrums get real," reads a 2015 headline in *Physics Today*, referring to new 3D printed eardrums with polymer scaffolds light and elastic enough to correct a perforation the way graft surgeries of small pieces of cotton wool or rubber failed to do. R. Mark Wilson, "Artificial Eardrums Get Real," *Physics Today* 68 (1 June 2015): 14.

100 Gisella Selden-Goth to Hallowell Davis, 2 Feb 1948, HDP box 12, folder 13.

101 George Wilson to E. E. Bullins, 9 February 1938, Albert D. Lasker Papers, box 11, folder 18. Huntington Library.

102 Hidden Hearing homepage, https://www.hiddenhearing.co.uk/hearing-aids /types-of-hearing-aids/spectacle-hearing-aids/.

Chapter 3

1   "The Deaf Hear: An Electrical Apparatus Causes Delight," *Boston Globe*, 29 August 1900, 12.

2   Lucy Taylor, "The Trials of the Partially Deaf," *The Silent Worker* 25.9 (June 1913): 163.

3   Bennett Chapple, "Curing the Deaf by Electricity," *National Magazine* (April 1903): 129–131.

4   Anna Wexler, "The Medical Battery in the United States (1870–1920): Electro-therapy at Home and in the Clinic," *Journal of the History of Medicine and Allied Sciences* 72.2 (2017): 172.

5   David E. Nye, *Electrifying America: Social Meanings of a New Technology, 1880– 1940* (Cambridge: MIT Press, 1990).

6   "They Give Hearing to Those Born Deaf," *Boston Journal* (28 August 1900), newspaper clipping, KBHA.

7   "The Deaf Hear."

8   Alexander L. Pach, "The Kinetoscope and New York Notes," *The Silent Worker* 14.7 (1902):102.

9   *The Canadian Mute*, reprinted in *The Silent Worker* 21 (June 1909): 174.

10 Kenneth Berger, *The Hearing Aid: Its Operation and Development* (Detroit: National Hearing Aid Society, 1970), 34.

11 R. E. Maynard, "For the Deaf and Dumb?" *The Silent Worker* 15.8 (1903): 123.

12 Taylor, "The Trials of the Partially Deaf," 163.

13 Axel Helmstädter, "Recharging the Battery of Life—Electricity in the Theory and Practice of Drug Treatment," *Pharmacy in History* 43.4 (2001): 141.

14 Paola Bertucci, "Shocking Subjects: Human Experiments and the Material Culture of Medical Electricity in Eighteenth-Century England," in Erika Dyck and Larry Steward (eds.), *The Use of Humans in Experiment: Perspectives from the 17th to the 20th Century* (Leiden: Brill, 2016), 112–113.

15 Paola Bertucci, "The Shocking Bag: Medical Electricity in Mid-18th-Century London," *Nuova Voltiana: Studies on Volta and His Times* 5 (2003): 31–42; François Zanetti, "Curing with Machines: Medical Electricity in Eighteenth-Century Paris," *Technology and Culture* 54 (2013): 503–530.

16 Iwan Rhys Morus, *Shocking Bodies: Life, Death, and Electricity in Victorian England* (Gloucestershire: History Press, 2011), 136.

17 Carolyn Thomas de la Peña, *The Body Electric: How Strange Machines Built the Modern American* (New York: New York University Press, 2003), 109.

18 J. T. H. Connor and Felicity Pope, "A Shocking Business: The Technology and Practice of Electrotherapeutics in Canada, 1849s to 1949s," *Material Culture Review* 49 (Spring 1999): 60–70.

19 Iwan Rhys Morus, "Marketing the Machine: The Construction of Electrothera-peutics as Viable Medicine in Early Victorian England," *Medical History* 36 (1992): 43.

20 Nancy Tomes, "Merchants of Health: Medicine and Consumer Culture in the United States, 1900–1940," *Journal of American History* 88.2 (2001): 533.

21   David G. Schuster, *Neurasthenic Nation: America's Search for Health, Happiness, and Comfort, 1869–1920* (New Brunswick, NJ: Rutgers University Press, 2011), 46.

22   Takahiro Ueyama, *Health in the Marketplace: Professionalism, Therapeutic Desires, and Medical Commodification in Late-Victorian London* (Palo Alto, CA: Society for the Promotion of Science and Scholarship, 2010), 114.

23   Jonathan Sterne, *The Audible Past: Cultural Origins of Sound Reproduction* (Durham, NC: Duke University Press, 2003), 81.

24   Gem Phone Co. pamphlet, *Cause Thine Ears to Hear* (1915), Trade catalog collection, Hagley Library and Archives.

25   Schuster, *Neurasthenic Nation*, 1.

26   Caroline Perkins to H. A. Tannous, 28 December 1897, Caroline White Siegfried & DeFois Hathaway Siegfried Archive Center, Rochester School for the Deaf.

27   "Deafness. How It Is Caused and How in Some Cases It Can Be Cured," *The Rochester Advocate* 8 (10 April 1888): 487–489.

28   Iwan Rhys Morus, "Batteries, Bodies, and Belts: Making Careers in Victorian Medical Electricity," in Paola Bertucci and Giuliano Pancaldi (eds.), *Electric Bodies: Episodes in the History of Medical Electricity* (Bologna: Università di Bologna, 2001), 212–213.

29   Morus, "Batteries, Bodies, and Belts," 214.

30   Morus, "Batteries, Bodies, and Belts," 229.

31   W. J. Tindall Co., "The Electricon—The Marvel of the Century," pamphlet, c. 1900, Bakken.

32   Oliver Hochadel, "'My Patient Told Me How to Do It': The Practice of Medical Electricity in the German Enlightenment," in Paola Bertucci and Giuliano Pancaldi (eds.), *Electric Bodies: Episodes in the History of Medical Electricity* (Bologna: Università di Bologna, 2001), 87.

33   Morus, "Marketing the Machine," 39.

34   Wexler, "The Medical Battery," 167.

35   Wexler, "The Medical Battery," 166.

36   Lori A. Loeb, "Consumerism and Commercial Electrotherapy: The Medical Battery Company in Nineteenth-Century London," *Journal of Victorian Culture* 4.2 (1999): 269; Ueyama, *Health in the Marketplace*, 21.

37   Ueyama, *Health in the Marketplace*, 152.

38   Samuel Hopkins Adams, *The Great American Fraud* (Chicago: Press of the American Medical Association, 1912), 106–108.

39   Pamphlets for these devices are contained in the "Deafness Cures" collection, HHF-AMA.

40   W. J. Tindall Co., "The Electricon—The Marvel of the Century," pamphlet, c. 1900, Bakken.

41   Advertisement for Dr. S. B. Smith's Torpedo Electro-Magnetic Machine, *Ladies' Wreath, a Magazine devoted to Literature, Industry, and Religion* (February 1850), 381.

42   Virginia P. Dean to G. M. Branaman, 10 July 1909, Rubenstein Rare Books and Manuscripts Library, Duke University, Eva Parrish Letters 1892–1909, box 164. On Branaman's company, see Christopher Hoolihan (ed.), *An Annotated Catalogue of the Edward C. Atwater Collection of American Popular Medicine and Health Reform*, vol. 3 (Rochester: University of Rochester Press, 2008), 90.

43   Daniel J. Robinson, "Mail-Order Doctors and Market Research, 1890–1930,"

in Hartmutt Berghoff, Phillip Scranton, and Uwe Spiekermann (eds.), *The Rise of Marketing Research* (New York: Palgrave Macmillan, 2012), 82.

44   Arthur J. Cramp, "Electric and Magnetic Cure-Alls," *Hygeia* (May 1939): 439–441, 462, 479.

45   Ueyama, *Health in the Marketplace*, 161.

46   Schuster, *Neurasthenic Nation.*

47   Tania Anne Woloshyn, *Soaking Up the Rays: Light Therapy and Visual Culture in Britain, c. 1890–1940* (Manchester: University of Manchester Press, 2017).

48   Tania Woloshyn, "'Kissed by the Sun': Tanning the Skin of the Sick with Light Therapeutics, c. 1890–1930," in Jonathan Reinarz and Kevin Patrick Siena, *A Medical History of the Skin: Scratching the Surface* (London: Routledge, 2016), 181.

49   "Be Healthy & Stay Healthy: Health Cure and Recovery by High-Frequency," unpublished manuscript (n.d.). Saint Louis Science Museum Collection.

50   Evan Yellon, *Surdus in Search of His Hearing: An Exposure of Aural Quacks and a Guide to Genuine Treatments and Remedies, Electric Aids, Lip-Reading and Employments for the Deaf Etc., Etc.* (London: Celtic Press, 1906).

51   Yellon, *Surdus in Search of His Hearing*, 41.

52   Yellon, *Surdus in Search of His Hearing*, 43.

53   Ueyama, *Health in the Marketplace*, 5.

54   Wexler, "The Medical Battery," 169.

55   *Healthy Rays from the Branston Violet Ray Generator*, pamphlet (1917), Electrotherapy, 1905–1974, box 229, folder 18, HHF-AMA.

56   Henry F. Garey, "The Efficacy of the Vibrometer in Applying Vibratory Massage in Aural Diseases," *Transactions of World Congress of Homeopathic Physicians and Surgeons* (May–June 1893): 434–443.

57   A. B. Norton, "The Vibrometer in Catarrhal Deafness," *Transactions of the Homeopathic Medical Society of the State of New York* 28 (1893): 339.

58   Joseph A. Danis to Arthur J. Cramp, 1 May 1924, Deafness Cures 1905–1973, box 181, folder 14, HHF-AMA.

59   "Dr. Guy Clifford Powell's Electro-Vibratory Apparatus for the Cure of Deafness," *The Lindan Collection of Medical Devices*, http://www.nolindan.com/lindan collection/ql208.html/.

60   Greg Fudacz, "Dr. Powell's Vibrator," *The Antikey Chop: Curios*, https://www .antikeychop.com/dr-powell-s-vibrator/.

61   Moon—or "Ellis," who was supposedly the inventor—did not advertise the Ear-O-Tone, but likely obtained names of hard of hearing persons and sent them form letters. Investigative inquiries by the Toledo League for the Hard of Hearing revealed that the company's offices, as listed in the letters, were unoccupied, and no contact information for the inventor was available. Ada Hill to Arthur J. Cramp, 2 November 1931; copy of a letter from Mrs. Todd Frendberg to Ada Hill, 14 April 1931, Deafness Cures 1905–1973, box 181, HHF-AMA.

62   "The Aural Vibrator," leaflet (n.d.); C. M. Jordan, "The Causes of Deafness and How Hearing Is Restored by the Aural Vibrator," pamphlet (n.d.), Medicine Series, box 5, folder 8, WCBA.

63   Robert Wendell Holmes III, "Substance of the Sun: The Cultural History of Radium Medicines in America," Ph.D. thesis, University of Texas at Austin, 2010.

64 Arthur J. Cramp, *Nostrums and Quackery and Pseudo-Medicine*, vol. 3 (Chicago: Press of the American Medical Association, 1936), 40.

65 Arthur J. Cramp to Elizabeth E. Sargent, 25 July 1929, Deafness Cures 1905–1973, box 181, folder 14, HHF-AMA.

66 M. M. Becker to AMA, 11 April 1931, Hearing Aids, 1907–1974, box 345, folder 5, HHF-AMA.

67 Arthur J. Cramp to Edmund Lissack, 28 February 1933, Hearing Aids, 1907–1974, box 345, folder 3, HHF-AMA.

68 Wexler, "The Medical Battery."

69 Lisa Rosner, "The Professional Context of Electrotherapeutics," *Journal of the History of Medicine and Allied Sciences* 43 (1988): 70.

70 Wexler, "The Medical Battery," 169.

71 Robert K. Waits, *The Medical Electricians: Dr. Scott and His Victorian Cohorts in Quackery* (Sunnyvale, CA: J.IV.IX Publications, 2013), 88.

72 Quoted in A. P. Ferry, "'Professor' William C. Wilson and His Actina Electric Pocket Battery for Curing Ocular Diseases," *Ophthalmology* 105.2 (1998): 244.

73 [New York & London Electric Association], "Positive Evidence that 'Actina' Cures," pamphlet, 1905, Bakken.

74 De la Peña, *The Body Electric*, 11.

75 C. C. Regier, "The Struggle for Federal Food and Drugs Legislation," *Law and Contemporary Problems* 1 (1933): 4.

76 Deborah Blum, *The Poison Squad: One Chemist's Single-Minded Crusade for Food Safety at the Turn of the Twentieth Century* (New York: Penguin Press, 2018).

77 Ferry, "'Professor' William C. Wilson and His Actina," 247.

78 *The Davis Electric Medical Battery: The New Way to Health*, pamphlet (1912), Electrotherapy, 1905–1974, box 229, folder 27, HHF-AMA.

79 "FDA's Evolving Regulatory Powers," U.S. Food and Drug Administration (1 February 2018), https://www.fda.gov/AboutFDA/History/FOrgsHistory /EvolvingPowers/default.htm/, accessed 4 December 2018.

80 Richard A. Merrill, "Regulation of Drugs and Devices: An Evolution," *Health Affairs* 10 (Summer 1994): 47–69.

81 Cramp, *Nostrums and Quackery*, vol. 3, 39.

82 "Deafness Cures," *Silent Worker* 34.4 (January 1922): 129.

83 Olive A. Whildin, "Cures for Deafness," *Silent Worker* 38.9 (June 1926): 426.

84 Oscar F. Swenson to Arthur J. Cramp, 28 December 1916, Deafness Cures 1905–1973, box 181, folder 14, HHF-AMA.

85 Elise D. Nelson to AMA, March 1933, Hearing Aids, 1907–1974, box 346, folder 2, HHF-AMA.

86 Berger, *The Hearing Aid*, 30–31.

87 Berger, *The Hearing Aid*, 38.

88 Memo, Arthur J. Cramp to R. R. Baer, 12 March 1918, Hearing Aids, 1907–1974, box 346, folder 2, HHF-AMA.

89 Arthur J. Cramp to G. H. Heald, 26 December 1928, Hearing Aids, 1907–1974, box 346, folder 2, HHF-AMA.

90 Akouphone Mfg. Co., booklet (n.d.), KBHA.

91 Acousticon Co., pamphlet (n.d.), KBHA.

92 Auto Ear Massage, pamphlet (n.d.), KBHA.

93   Swenson to Cramp, 28 December 1915.

94   G. H. Heald to Arthur J. Cramp, 23 December 1928, Hearing Aids, 1907–1974, box 346, folder 2, HHF-AMA.

95   *The Telonor With the Sensitone*, pamphlet (1916), Hearing Aids, 1907–1974, box 327, folder 1, HHF-AMA.

96   Mears Ear Phone Co., form letter, 23 February 1927, Hearing Aids, 1907–1974, box 346, folder 2, HHF-AMA.

97   Annetta W. Peck to Arthur J. Cramp, 26 September 1930, Hearing Aids, 1907–1974, box 346, folder 2, HHF-AMA.

98   Memo from Federal Trade Commission, 2 February 1934. Hearing Aids, 1907–1974, box 346, folder 2, HHF-AMA.

99   Merrill, "Regulation of Drugs and Devices," 55.

100 Ueyama, *Health in the Marketplace*, 15.

Chapter 4

1   "Flying Cures Deafness," *Stanberry Headlight* (Stanberry, MO), 8 September 1927, 6.

2   Frances Warfield, "I Was an Exceptional Child," *Exceptional Children* 19.1 (October 1952): 3.

3   "Regains hearing in Plane Flight: Ex-Service Man Hears After Test Here," *Chicago Tribune*, 4 September 1922, 1.

4   "9000-Foot Aerial Dives Taken by Oakland Pianist To Cure Deafness," *Oakland Tribune*, 11 May 1925, 1.

5   "Plane Rides with Dives at High Altitudes Tried to Cure Deafness," *Hutchinson News*, 2 September 1929.

6   "Flying for Deafness," *The Times Herald* (Port Huron, MI), 10 June 1931, 6.

7   "Doctor Sends Deaf Dog Aloft for 'Fright' Cure; Thaw's Son Goes Along," *The Brooklyn Daily Eagle* 21 August 1928, 3.

8   "Girl Mute Tries Flight As Cure," *The Brooklyn Daily Eagle*, 21 September 1928, 3.

9   "Woman, Child Try Plane Drop to Cure Deafness," *Oakland Tribune* 21 August 1928.

10   "The Airplane Cure," *Wilmington News-Journal*, 14 September 1925.

11   "Benefits of Airplane Nose Diving for Deafness Doubtful," *Oakland Tribune*, 12 October 1925, 1.

12   A selection of stories: "Stunt Flying Cost Three Lives," *Scranton Republican*, 6 September 1928, 1; "Deaf, Dumb for 19 Years, Flight In Air Aids Cure," *Chicago Tribune*, 11 December 1924, 16; "Deaf Boy, Taking Flying Cure, Killed With Pilot," *Chicago Tribune*, 5 September 1925, 2; "Airman Flies to Cure Deaf Boy; Both Killed," *Oakland Tribune*, 4 September 1925, 20; "Deaf Pianist Killed in Plane," *Oakland Tribune*, 30 April 1928, 3; "Seeking Deaf Cure Dies With Aviator," *The Scranton Republican*, 2 November 1925, 1.

13   "Mahan, Deaf Boxer, Dies in a Parachute Leap of 5,000 Feet from Plane to Regain Hearing," *New York Times*, 24 February 1930.

14   "'Airplane Cures' for the Deaf a Myth," *The Silent Worker* 41.4 (April 1929): 136–137.

15   *The California News*, reprinted in *Rochester Advocate* 46 (January 1926): 9.

16 Arthur Cramp to J. A. Deegan, 12 January 1931, Deafness Cures 1905–1973, box 182, folder 1, HHF-AMA.

17 "Flying for Deafness," *Brooklyn Daily Eagle*, 17 October 1928, 8.

18 George B. McAuliffe, "New Help for the Hard of Hearing," *Popular Science Monthly* (February 1930): 67–68, 141.

19 "Hold Stunt Flying No Deafness Cure," *Brooklyn Daily Eagle*, 26 October 1932, 20.

20 Claire Badaracco, *Prescribing Faith: Medicine, Media, and Religion in American Culture* (Waco: Baylor University Press, 2007), 4.

21 Joel Paris, *Fads and Fallacies in Psychiatry* (London: Royal College of Physicians, 2013).

22 Peter N. Stearns, *Fat History: Bodies and Beauty in the Modern World* (New York: New York University Press, 1997), 25; Andrew McClary, "What Is a Health Fad? The Posture Movement as an Example," *Journal of American Culture* 6.4 (1983): 50–55.

23 Holly Folk, *The Religion of Chiropractic: Populist Healing from the American Heartland* (Chapel Hill: University of North Carolina Press, 2017), 19; Harry B. Weiss and Howard R. Kemble, *The Great American Water-Cure Craze: A History of Hydropathy in the United States* (Trenton, NJ: Past Times Press, 1967).

24 Folk, *The Religion of Chiropractic*, 2.

25 Eric S. Juhnke, *Quacks and Crusaders: The Fabulous Careers of John Brinkley, Norman Baker, and Harry Hoxsey* (Lawrence: University Press of Kansas, 2002).

26 John C. Burnham, *Health Care in America: A History* (Baltimore: Johns Hopkins University Press, 2015), 387.

27 Dale C. Smith, "Appendicitis, Appendectomy, and the Surgeon," *Bulletin of the History of Medicine* 70.3 (1996): 414–441.

28 Paris, *Fads and Fallacies in Psychiatry*, 21.

29 George E. Shambaugh, "Fads and Fancies in the Practice of Otolaryngology," *Journal of the American Medical Association* 87.21 (20 November 1926): 1720.

30 McClary, "What Is a Health Fad?" 51.

31 Louis Dwyer-Hemmings, "'A Wicked Operation'? Tonsillectomy in Twentieth-Century Britain," *Medical History* 62.2 (2018): 217–241.

32 Lawrence D. Longo, "The Rise and Fall of Battey's Operation: A Fashion in Surgery," *Bulletin of the History of Medicine* 53.2 (1979): 244–267. Longo argues that perhaps the most important factor leading to the operation's demise was its "almost wholesale use for the cure of convulsive disorders and insanity," including nymphomania and other mental disorders.

33 Baron H. Lerner, *The Breast Cancer Wars: Hope, Fear, and the Pursuit of a Cure in Twentieth-Century America* (New York: Oxford University Press, 2001); David S. Jones, "Surgery and Clinical Trials: The History and Controversies of Surgical Evidence," in *The Palgrave Handbook of the History of Surgery*, ed. Thomas Schlich (London: Palgrave Macmillan, 2016), 479–501.

34 Erika Janik, *Marketplace of the Marvelous: The Strange Origins of Modern Medicine* (Boston: Beacon Press, 2014); Emily Odgen, *Credulity: A Cultural History of US Mesmerism* (Chicago: University of Chicago Press, 2018).

35 James C. Whorton, *Nature Cures: The History of Alternative Medicine in America* (Oxford: Oxford University Press, 2002), 166–167.

36 Whorton, *Nature Cures*, 168; Simon A. Senzon, "Constructing a Philosophy of

Chiropractic: Evolving Worldviews and Postmodern Core," *Journal of Chiropractic Humanities* 18.1 (December 2011): 39-63.

37 Mrs. Joseph Rosette, "The Deaf Hear," *The Magnetic Cure*, no. 15 (January 1896), 3.

38 *Davenport Leader* (13 May 1894), 5.

39 D. D. Palmer, *The Chiropractor's Adjuster* (Portland, OR: Portland Printing House Company, 1910), 18.

40 Harvey Lillard, "Deaf Seventeen Years," *The Chiropractic*, no. 17 (January 1897): 3.

41 Folk, *The Religion of Chiropractic*, 19.

42 Three issues of *The Chiropractic* survive from 1899-1902. As Holly Folk points out, it appears most of Palmer's work was redundant as he reprinted similar materials frequently. Folk, *The Religion of Chiropractic*, 95.

43 "Deafness Can Be Cured," *The Chiropractic*, no. 26 (November 1899), 4. Special thanks to Rosemary Riess for drawing my attention to this letter.

44 Paul Benedetti and Wayne MacPhail, *Spin Doctors: The Chiropractic Industry Under Examination* (Toronto: Dundurn Press, 2002).

45 D. D. Palmer, *The Chiropractor* (Los Angles: Press of Beacon Light Printing Company, 1914).

46 Folk, *The Religion of Chiropractic*, 95.

47 Benedetti and MacPhail, *Spin Doctors*, 36.

48 Folk, *The Religion of Chiropractic*, 18.

49 Folk, *The Religion of Chiropractic*, 100-102.

50 Rolf E. Peters and Mary Ann Chance, "Murder They Wrote: The Death of D. D. Palmer and Its Aftermath," *Chiropractic Journal of Australia* 23.4 (December 1993): 143-148.

51 Due to his deafness, in 1933 the prince surrendered his rights to the line of succession and of his future heirs, a decision that would have profound ramifications for Spanish politics thirty years later, when he recanted his renunciation. His decision to recant was made after his second wife, German singer Charlotte Teidemann, helped him improve his speech. See Charles Powell, *Juan Carlos of Spain: Self-Made Monarch* (London: Macmillan Press, 1996).

52 "Extraordinary Discovery About the Queen of Spain's Deaf Son," *Great Falls Daily Leader*, 7 August 1920, 7; "Don Jaime's Deafness," *Chicago News*, 17 October 1923, newspaper clipping, Deafness Cures 1905-1973, box 183, folder 6, HHF-AMA; "Spanish King's Son Patient of Muncie," unknown Brooklyn paper, 8 July 1923, newspaper clipping, Department of Investigation Records, box 519, folder 8, HHF-AMA.

53 "The Ambassador said that he knew nothing positive about the alleged cure of Prince Don Jaime, but that, personally, he doubted it very much; if it were true, Spanish papers would have mentioned it, in view of the general interest in everything pertaining to the royal family. He added that Don Jaime had been seen by the best physicians, who had all agreed that his trouble was incurable, as the auditory nerves were not atrophied but destroyed," confidential statement of the Spanish ambassador, 6 October 1920, Deafness Cures 1905-1973, box 183, folder 6, HHF-AMA.

54 Curtis H. Muncie to *Volta Review*, 25 November 1923, Department of Investigation Records, box 519, folder 9, HHF-AMA.

55 Ravi S. Swamy and Robert K. Jackler, "The Fickle Finger of Quackery in Otology:

The Saga of Curtis H. Muncie, Osteopath," *Otology and Neurotology* 31 (2010): 846–855. Thanks to Robert K. Jackler for drawing my attention to the allegedly soundproof room.

56 Edgar S. Kennedy, "Has Most Valuable Hands in the World," *Sunday Eagle Magazine*, 14 February 1926, 3; "Skilled Fingers Used in Knifeless Surgery," *New York City News*, 16 November 1924, newspaper clipping, Department of Investigation Records, box 519, folder 8, HHF-AMA.

57 Invoice for examination Curtis H. Muncie address to redacted name, 9 February 1925; Curtis H. Muncie to Katie L. Baker, 12 September 1923, Department of Investigation Records, box 519, folder 9, HHF-AMA.

58 "Crowds Wait for 'Finger Surgeon' to Heal Deafness," *Buffalo New York Courier* (6 April 1923), newspaper clipping, Department of Investigation Records, box 519, folder 8, HHF-AMA.

59 On orificial surgery, see Natasha Frost, "Orificial Surgery Was a Gruesome 19th-Century Sham Medicine," *Atlas Obscura* (16 January 2018): https://www .atlasobscura.com/articles/sham-medicine-orificial-surgery-edwin-pratt-vagina -phrenology/.

60 In their attempts to secure monopoly and define professional standards, the orthodox medical establishment attempted to outlaw competition from groups that they believed threatened the therapeutic consensus of the profession— "cults" as they were called—including osteopathy. Despite legislative attempts to limit the autonomy of osteopaths, by the 1930s osteopathy solidified as a distinct field with its own institutions, journals, and licensing boards. For a comprehensive history of osteopathy, see Norman Gevitz, *The Dos: Osteopathic Medicine in America* (Baltimore: Johns Hopkins University Press, 2004). Furthermore, by keeping the nose and throat clean and teeth in good condition, and encouraging breathing pure outdoor air, osteopathy was considered a revolutionary treatment for avoiding head colds that could cause hearing loss. One recommendation was to sleep outdoors as the best way to achieve this. "How to Keep Fit—Deafness Is Preventable," *Hutchinson Blade*, 9 October 1930, 1.

61 Curtis H. Muncie, *Prevention and Cure of Deafness through Muncie Reconstruction Method* (New York: Curtis H. Muncie, 1936), 52, Department of Investigation Records, box 519, folder 7, HHF-AMA. The pamphlet was likely first printed in 1920, as the second edition was released in 1921. Subsequent editions were published in 1924, 1936, 1941, 1948, 1957, and 1960.

62 "Finger Surgery as a Cure for Deafness," *American Annals of the Deaf* 68, no. 4 (September 1923): 347.

63 Swamy and Jackler, "The Fickle Finger of Quackery," 848.

64 "Cure for Deaf in 10 Minutes! Tiny Operation," *Chicago Herald & Examiner*, 14 May 1936, newspaper clipping, Department of Investigation Records, box 519, folder 8, HHF-AMA.

65 Viola Roseboro, "New Successes in Treating Deafness," *McClure's Magazine* (May 1925), 113–124.

66 William H. Walsh to Mr. Sanger, 4 June 1925, Department of Investigation Records, box 519, folder 9, HHF-AMA.

67 E. J. Goodwin to Arthur J. Cramp, 2 July 1925, Department of Investigation Records, box 519, folder 9, HHF-AMA.

68 George E. Shambaugh to Morris Fishbhein, 16 February 1926; George E. Sham-

baugh to S. S. McClure, 16 February 1926, Department of Investigation Records, box 519, folder 9, HHF-AMA.

69 Winifred H. Blanchard to Miss Murphy [19 June 1926], Department of Investigation Records, box 519, folder 9, HHF-AMA.

70 Mrs. H. Hallenstein to editor, *JAMA* 13 April 1926, Department of Investigation Records, box 519, folder 9, HHF-AMA.

71 *Hygeia* (February 1924), 118.

72 W.H.B., "Finger Surgery," *Hygeia*, 11 June 1923, 327.

73 Charles A. Croissant to AMA, 14 September 1925, Department of Investigation Records, box 519, folder 9, HHF-AMA.

74 C. H. Ames to editor of *Hygeia*, 19 February 1926, Department of Investigation Records, box 519, folder 9, HHF-AMA.

75 L. M. Hanks to Josephine B. Timberlake, 2 May 1926, Department of Investigation Records, box 519, folder 9, HHF-AMA.

76 "U.S. Finds 'Finger Surgeon' Fraud in Income Tax," *Journal of the American Medical Association* 119.10 (4 July 1942), 799.

77 "'Finger Surgery' for Deaf Children," *Hearing Dealer* (February 1953): 6–7. Much to Yale otolaryngologist Norton Canfield's disdain, the magazine identified Douglas J. Muncie as "Chief Consultant in Audiology Department of Medicine and Surgery Veterans Administration"—a glaring error, given that that was Canfield's title and that his article in the issue immediately followed Muncie's! Canfield, box 11, folder, "Audiology Correspondence 1949–70."

78 Letter from Hallowell Davis to R. J. Marquis, 19 January 1959, HDP box 23, folder 11.

79 J. D. Ratcliff, "A Window for Deaf Ears," *Hearing Dealer* 2.8 (August 1952): 8–9.

80 Howard P. House, "What Can Be Done for the Deafened Today," *California and Western Medicine* 63.3 (September 1945): 130–132.

81 Neil Weir, *Otolaryngology: An Illustrated History* (London: Butterworths, 1990), 203.

82 Gösta F. Dohlman, "Carl Olof Nylén and the Birth of the Otomicroscope and Microsurgery," *Archives of Otolaryngology* 90.6 (1969): 813–817.

83 Julius Lempert, "An Analytical Study of the Evolutionary Development of the Fenestration Operation," *Annals of Otology, Rhinology, and Laryngology* 59.4 (December 1950): 988–1019.

84 Norton Canfield to Roderick Hefron, 8 June 1938. Canfield, box 7.

85 Terence Cawthorne, "Julius Lempert: A Personal Appreciation," *Archives of Otolaryntology* 90 (December 1969): 680–686.

86 Gordon D. Hoople, "Personal Recollections of Julius Lempert," *Archives of Otolaryntology* 90 (December 1969): 690–693.

87 George E. Shambaugh Jr., "Hearing Results in 2,100 Consecutive Fenestration Operations During Ten Years," *Acta Oto-Laryngologica* 37, Supplementum 79 (1949): 85–101.

88 Ivan Illich, *Medical Nemesis: The Expropriation of Health* (London: Calder & Boyars, 1975).

89 [Emanuel M. Josephson], *The Lempert Fenestration Operation for Deafness: Mayhem and Human Experimentation* (n.d.), Deafness Cures, 1905–1973, box 182, folder 20, HHF-AMA. Reprinted in Emanuel M. Josephson, *Your Life Is Their Toy: Merchants in Medicine* (New York: Chedney Press, 1948).

90 Oliver Field to Dr. X. Corso, 11 May 1950, Deafness Cures, 1905–1973, box 182, folder 20, HHF-AMA.

91 J. D. Ratcliff, "A Window for Deaf Ears," *Hearing Dealer* (December 1952), 8.

92 "New Window In Ear Made Permanent So Deaf Can Hear," *New York World*, telegram, 21 March 1940. Deafness Cures 1905–1973, box 182, folder 20, HHF-AMA.

93 Damon Runyon, "Dr. Julius Lempert—A Great Man," *New York Journal*, 4 August 1945.

94 *Time* (18 January 1960). Margaret Sullavan suffered hearing impairment for seventeen years; Lempert successfully restored hearing in her left ear, but her right ear gradually worsened. After Sullivan died from an overdose of sleeping pills, it was revealed that she willed her temporal bones to Lempert for scientific study, as part of what became the Deafness Research Foundation, a temporal bone bank that was initially established thanks to a generous donation from Leonard Firestone. "Doctor Willed Ear Bones by Miss Sullavan," *Chicago Daily Tribune*, 9 January 1960, press clipping, Deafness Cures 1905–1973, box 182, folder 20, HHF-AMA.

95 "Physician Sees Great Future in Ear Surgery," news clipping (n.d.), Deafness Cures 1905–1973, box 182, folder 20, HHF-AMA.

96 "Facts about Fenestration," *Bulletin of the New York League for the Hard of Hearing* 23.2 (April 1945).

97 C. Stewart Nash, "Fenestration Operation—Its Pros and Cons," *Hearing News* 16.3 (March 1948): 1–4.

98 Fay Sobel to Hallowell Davis, 10 November 1946, S—Miscellaneous, 1937–1948, box 4, folder 12, HDP.

99 *Hygeia*, 25 September 1940.

100 "$24,000 Verdict Against Surgeon," press clipping, 27 January 1944, Deafness Cures 1905–1973, box 182, folder 20, HHF-AMA.

101 I. Smison Hall, "The Fenestration Operation for Otosclerosis," *British Medical Journal*, no. 4478 (2 November 1946): 648. Hall added that in patients over forty years of age, there was a higher risk of nerve degeneration that would prevent hearing improvement after operation.

102 George Shambaugh Jr. noted that "since the proportion of otosclerosis to the total number of hard of hearing has . . . been estimated from 5% to 10% this would indicate that only 1.5% to 5% of the hard of hearing population would be good candidates for the fenestration operation." George Shambaugh Jr., quoted in Leland A. Watson and Thomas Tolan, *Hearing Tests and Hearing Instruments* (Baltimore: Williams & Wilkins Co., 1949), 428.

103 Watson and Tolan, *Hearing Tests and Hearing Instruments*, 429.

104 Letter from F. Reinoso to Hallowell Davis, 16 November 1959, HDP box 24, folder 8.

105 *Endicott Daily Bulletin*, 5 August 1943, 12.

106 Theodore E. Walsh to Hallowell Davis, 22 September 1951, HDP box 15, folder 26.

107 Report by George E. Shambaugh Jr., Meeting of Otosclerosis Surgeons, 20 October 1941, Canfield, box 7.

108 George Shambaugh Jr., "Julius Lempert and the Fenestration Operation," *American Journal of Otology* 16.2 (March 1992): 249.

109 Dragan Dankuc, "History of the Surgery for Otosclerosis and Cochlear Implants," *Medicinski Pregled* 5–6 (2015): 151.

110 Gunnar Holmgren to Norton Canfield, 29 May 1946, Canfield, box 11.
111 Cawthorne, "Julius Lempert: A Personal Appreciation." Shambaugh testified that Lempert "epitomized the American tradition of the self-made man," rising from rags to riches. Lempert achieved remarkable successes: the Brazilian government bestowed upon him their highest honor, the Chevalier of the Southern Cross; he became the second American (after Harvey Cushing) and first otologist to be awarded an honorary M.D. degree from Stockholm's Karolinska Institute; and he received numerous other awards and accolades for the fenestration operation. In May 1948, the Otological Section of the Royal Society of Medicine invited him to read a paper and exhibit his work at the West Hall of the Royal College of Medicine. With eight panels of enlarged photomicrographs of temporal bone pathology and a dozen dissected heads, Lempert demonstrated to his professional brethren his qualifications, worth, and dignity. In one instance, Lempert invited a hundred leading American otologists to New York City for a magnificent dinner to celebrate Holmgren when he visited the United States. The dinners eventually became customary; Lempert and his wife hosted them following annual meetings of the American Academy of Ophthalmology and Otolaryngology, usually at the Pump Room in Chicago. "To be invited to one of these dinners," recalled Cawthorne, "was to be regarded as one of the elite in otolaryngology."
112 John E. Vacha, "Fads and Fancies," *Encyclopedia of Cleveland History* (2018), https://case.edu/ech/articles/f/fads-and-fancies/.
113 Dmitriy Niyazov, Alexandra Borges, Akira Ishiyama, Edward Zaragoza, and Robert Lufkin, "Fenestration Surgery for Otosclerosis: CT Findings of an Old Surgical Procedure," *American Journal of Neuroradiology* 21 (October 2000): 1670–1672; Millicent King Channell, "Modified Muncie Technique: Osteopathic Manipulation for Eustachian Tube Dysfunction and Illustrative Report of Case," *Journal of the American Osteopathic Association* 108.5 (May 2008): 260–263; Joseph O. Di Duro, "Improvement in Hearing After Chiropractic Care: A Case Series," *Chiropractic and Osteopathy* 14.2 (2006): 1–7.
114 *The Silent Worker* 15.4 (January 1913): 70.
115 Letter from Ethel F. Butts to Hallowell Davis, [November 1959], HDP box 22 folder 1.

Chapter 5

1 "The World's Most Attractive Hearing Aid," transcript for sound film. Zenith Hearing Aid Dealer Manual, Volume 1: "How to Sell Your Product" (1957). Manuscript, KBMA.
2 U.S. Senate Hearings, "Hearing Loss, Hearing Aids, and the Elderly," Subcommittee on Consumer Interests of the Elderly, Special Committee on Aging, U.S. Senate, 90th Congress, 2nd Session, 18 and 19 July 1968, 61.
3 Cecile Starr, "Ideas on Film: Free Films Round-Up," *Saturday Review* (11 August 1956), 38.
4 *Journal of the American Medical Association* (22 December 1956), 1569.
5 Zenith Hearing Aid Dealer Manual Volume 1, "How to SELL Your Product," KBHA.
6 Arthur Hambleton to Howard A. Carter, 27 September 1951, HDP box 11, folder 25.

7   "Speaking of Appearances," *Better Living Magazine* (February 1939), 10.

8   Pamphlet, *A Miracle of Modern Happiness* (n.d.), KBHA.

9   Kenneth Berger, *The Hearing Aid: Its Operation and Development* (Detroit, MI: National Hearing Aid Society, 1970), 53.

10 A Miracle of Modern Happiness *A Miracle of Modern Happiness* (n.d.), KBHA.

11 None of these men was in the hearing aid business before establishing National Electronics, nor did they stay in the industry after voluntarily dissolving the company and selling it to another dealer in Gloversville. "This was not a serious effort on our part," Hutchings later recalled, "just a sideline as all 3 of us were [General Electric] engineers." Note by John H. Hutchings in a letter addressed by Kenneth Berger, KBHA.

12 Acousticon, for instance, established an academy for training their "army of Acousticians" (as their dealers were known) in the technical nature of the company's products, and also in the rudiments of ear anatomy, the diseases and physiology of deafness, audiometry, and the psychological nature of hearing impairment, including the complications that poor hearing can precipitate for the nervous system. This training process went beyond innovation, risk, or even ownership of products, instead underscoring the necessity for dealers to foster relationships with prospects.

13 Acousticon Sales Training Manual, c. 1950s, KBHA.

14 "1952: Transistorized Consumer Products Appear," The Computer History Museum, https://www.computerhistory.org/siliconengine/transistorized-consumer-products-appear/.

15 "1952: Transistorized Consumer Products Appear," The Computer History Museum, https://www.computerhistory.org/siliconengine/transistorized-consumer-products-appear/.

16 Zenith advertisement, "Now, with her Zenith Hearing Aid, Mother Can Hear as well as Dad and Sonny!" *Time* (29 October 1951): 13, JHC.

17 The model also relied upon a contralateral routing of sound (CROS), in which sound from the worsened ear could be transmitted from the stronger ear. Neil Bauman, "CROS Hearing Aids Existed Before CROS Hearing Aids Were Invented," *Canadian Audiologist* 3.1 (2016): http://canadianaudiologist.ca/2746-feature/.

18 Bauman, "CROS Hearing Aids."

19 Hallowell Davis to Arthur M. Churchill, 19 November 1952, HDP, box 12, folder 2.

20 Zenith Radio Co., *Frauds and Facts—Some Things That Go On in the Sale of Hearing Aids* (Chicago, Zenith Radio Corp., c. 1950), 3.

21 The rationing of batteries, moreover, was a source of constant anxiety for wearers. The New York League for the Hard of Hearing, for instance, worked with otologists and the industry to ensure that batteries were available for users; indeed, the shortage led to further development of batteries, including the Power-Pill. Acousticon advertisement, *Toledo Times* (21 July 1858): "Tiny Power Pill: answer to your dream of miracle hearing. . . . incredible featherweight is only 2.2 ounces and most vital of all, it has been designed to bring you what we believe is the closest thing to natural hearing," KBHA.

22 For a historical overview of the Zenith Radio Company, see Harold N. Cones and John H. Bryant, *Zenith Radio: The Glory Years 1936–1945* (Atglen, PA: Schiffer Publishing Ltd., 2003).

23 Zenith advertisement, "You owe it to your Uncle Sam," *Popular Science* 114 (February 1944): 225.

24 Zenith advertisement. "To the Readers of *Hygeia*," *Esquire* (February 1944): 85, JHC.

25 Rupert Hughes, *How Benjamin Franklin Helped Bring Better Hearing to Modern Americans* (Chicago: Zenith Radio Corp., 1944), Franklin Collection, Sterling Memorial Library, Yale University.

26 Zenith advertisement, "Deaf Bob," *Hygeia* (September 1945): 689, JHC.

27 Sonotone advertisement, *Life* (27 September 1943).

28 Mara Mills, "Hearing Aids and the History of Electronics Miniaturization," *IEEE Annals of the History of Computing* 33.2 (April–June 2011): 24–45.

29 Acousticon advertisement, c. 1959, Medicine Series, box 5, folder 3–4, WCBA.

30 Beltone Hearing Aid Co., "A New Atomic Age Miracle and What It Means to Your Hearing!" pamphlet, c. 1950s, Medicine Series, box 5, folder 10, WCBA. See also "A Modern Arabian Nights Story: 'The Miracle of Better Hearing," advertisement leaflet, 1953, Hearing Aids, 1907–1974, box 345, folder 3, HHF-AMA.

31 Zenith advertisement, "To the Readers of *Hygeia*," *Esquire* (February 1944): 85, JHC; Anne Wheeler to Steven M. Spencer [September 1946], Canfield, box 7.

32 "In this business," the operating manual for Acousticon Managers and Distributors stated, "the names and addresses of 1,000 hard-of-hearing people—users and non-users—are more valuable than $1,000 in cash or merchandise." "How to Find, Hire, Train, and Manage Acousticians: An Operating Manual for Acousticon Managers and Distributors," KBHA.

33 Circular, American Ear Phone Company (c. 1938). Mary Lee of Chestnut Hill, Massachusetts, an Audi-Ear wearer, recommended a person to the company for commission. She was additionally advised to persuade the new customer to keep the hearing aid until the trial period was over, and then she would receive the commission. Papers of Mary Lee, Schlesinger Library, Harvard University.

34 John H. Fredrickson to Steven M. Spencer, 20 September 1946. Canfield, box 7.

35 Fred W. Kranz to Hallowell Davis, 17 December 1948, RG032 Series 3: Hearing Aid Device Manufacturers, Literature and Correspondence, box 17, CID.

36 Hallowell Davis to Fred W. Kranz, 20 December 1948, RG032 Series 3: Hearing Aid Device Manufacturers, Literature and Correspondence, box 17, CID.

37 Papers of Theresa Goell, 1906–2005, MC 534. box 16, folder 14; box 17, folder 2, Schlesinger Library on the History of Women in America, Radcliffe Institute for Advanced Study (hereafter "PTG").

38 Theresa Goell to William Röwlönd, 15 June 1955, Series II, Biographical and Personal; Subseries C, Medical, 1939–1986, Sonotone Corporation, PTG.

39 Theresa Goell to S. F. Morgan, 16 April 1958, Series II, Biographical and Personal; Subseries C, Medical, 1939–1986, Sonotone Corporation, PTG.

40 Anna G. Creadick, *Perfectly Average: The Pursuit of Normality in Postwar America* (Amherst: University of Massachusetts Press, 2010), 143.

41 Hillel Schwartz, "Hearing Aids: Sweet Nothings, or an Ear for an Ear," in Pat Kirkham (ed.), *The Gendered Object* (Manchester: Manchester University Press, 1996), 55.

42 *Better Living* (Autumn 1937), Library of Congress collection. Sonotone launched the magazine after deciding to fold its *Hearing News Bulletin*, an annual newsletter that was appreciated by their customers, but also limited in scope, as it did

not reach anywhere close to the nearly five million hearing-impaired persons in the United States. The company thus designed their new magazine to fill a need for authentic and helpful information on various aspects of hearing problems to attract potential readers and prospects and provide them with facts concerning deafness.

43 Creadick, *Perfectly Average*, 120.

44 Sonotone, "Fashion: Your Passport to Poise," pamphlet, 1950, RG032 Series 3: Hearing Aid Device Manufacturers, Literature and Correspondence, box 17, CID.

45 "I am deaf but not sorry," magazine clipping, Canfield, box 11.

46 *Aladdin Brings New Horizons to Hearing*, pamphlet (c. 1945), KBHA.

47 *The Silent World* (March 1947), 280.

48 *The Silent World* (September 1948), 123.

49 Acousticon advertisement in *Toledo Blade* (2 September 1951), clipping, *Toledo Blade*, KBHA.

50 Acousticon advertisement in *Toledo Blade* (30 September 1951), clipping, *Toledo Blade*, KBHA.

51 Advertisement for Acousticon DR-1 Special model, *Toledo Blade* (30 September 1951), clipping, *Toledo Blade*, KBHA.

52 Tonemaster advert, *Hearing Dealer* (May 1956); photograph of Soundfinder "Ear Ring type," c. 1965, developed by Starkley Labs, Minneapolis, KBHA.

53 Acousticon advertisement in *Toledo Blade* (28 August 1951), clipping, *Toledo Blade*, KBHA.

54 John S. Meagher to Steven M. Spencer, 3 September 1947, Canfield, box 7.

55 Steven M. Spencer to Norton Canfield, 25 September 1946, Canfield, box 7.

56 "Speaking of Appearances," *Better Living Magazine* (February 1939), 10.

57 Paravox internal memo, "The HEAR-Zone," 28 January 1948, RG032 Series 3: Hearing Aid Device Manufacturers, Literature and Correspondence, box 17, CID.

58 Paravox advertisement, c. 1948, RG032 Series 3: Hearing Aid Device Manufacturers, Literature and Correspondence, box 17, CID.

59 Dictograph Products Inc., "The Scientific Correction of Hearing by Acousticon," internal dealer booklet (5 July 1957), KBMA.

60 *The Scientific Correction of Hearing.*

61 Edward P. Fowler Sr. to Irving Schachtel, 11 June 1952, HDP box 11, folder 25.

62 Rosemary Stevens, *American Medicine and the Public Interest* (Berkeley: University of California Press, 1998), 157.

63 Draft report, Advisory Committee on Audiometers and Hearing Aids, 15 January 1951, Change of Policy RE: Advertising of Hearing Aids, 1950–1950, box 9, folder 7, HDP.

64 "Hearing Aid Advertising Recommendations," National Better Business Bureau Report, July 1951, HDP box 13, folder 21.

65 Of course, some manufacturers insisted AMA guidelines needed to be adhered to, especially when it gave them an added advantage when the AMA penalized competitors. Leland A. Watson, president of Maico, for instance, wrote to council secretary Howard A. Carter "Mr. McDonald and the Zenith Corporation would never have anything to do in its sales program with hearing aid dealers as such, yet in the past year the Zenith Corporation has affiliated itself with some of the worst elements in the hearing aid industry, some of those who have been most frequently cited by the local Better Business Bureaus for unethical advertising

and unethical practices, and some of these same individuals continue to carry the Zenith flag and advertising statements with impunity which are in direct violation of the A.M.A. requirements." Leland A. Watson to Howard A. Carter, 4 January 1951, HDP box 9, folder 7.

66 "Advisory Committee on Audiometers and Hearing Aids, Communication #125, 15 January 1951," HDP box 9, folder 7.

67 Chicago Society for the Hard of Hearing Report for Regular Board Meeting, 14 December 1944. Chicago Society for the Hard of Hearing Papers, Chicago History Museum.

68 Howard A. Carter, "Hearing Aids and Advertising," *Journal of the American Medical Association* 146.7 (30 June 1951): 651.

69 Carter, "Hearing Aids and Advertising," 651.

70 Howard A. Carter to Hallowell Davis, 13 April 1949, Council on Physical Medicine—Audiometers and Hearing Aids, 1949, box 2, folder 5, HDP.

71 Hallowell Davis to Howard A. Carter, 9 February 1949, Council on Physical Medicine—Audiometers and Hearing Aids, 1949, box 2, folder 5, HDP.

72 Memo from M. S. Steenson to distributors, 3 June 1949, Dictograph Products Statement Report, 31 August 1949, Council on Physical Medicine—Audiometers and Hearing Aids, 1949, box 2, folder 5, HDP.

73 Harry G. Lang and Bonnie Meaath-Lang, *Deaf Persons in the Arts and Sciences: A Biographical Dictionary* (Westport, CT: Greenwood Press, 1995), 118-120; L. A. Davies, "Successful Deaf People of Today: Dorothy Canfield Fisher," *Volta Review* 24.5 (1922): 148-150.

74 Fred J. Wonders to Howard A. Carter, 1 September 1949, RG032 Series 3: Hearing Aid Device Manufacturers, Literature and Correspondence, box 17, CID.

75 Fred J. Wonders, "Paravox Fall 1949: Advertising Publicity," internal memo booklet (1 September 1949), RG032 Series 3: Hearing Aid Device Manufacturers, Literature and Correspondence, box 17, CID. (for reference to Paravox advertising strategy on celebrities).

76 Howard A. Carter to Hallowell Davis, 1 March 1948, Council on Physical Medicine—Audiometers and Hearing Aids, 1946-48, box 2, folder 4, HDP.

77 Irving Schachtel to Hallowell Davis, 4 June 1952, General Correspondence, 1951-1952, box 11, folder 25, HDP.

78 Leland A. Watson to Howard A. Carter, 31 July 1952, General Correspondence, 1951-1952, box 11, folder 25, HDP.

79 Howard A. Carter to Irving I. Schachtel, 17 February 1948, Council on Physical Medicine—Audiometers and Hearing Aids, 1946-48, box 2, folder 4, HDP.

80 W. N. Brown to Ralph E. DeForest, 2 May 1952, RG032 Series 3: Hearing Aid Device Manufacturers, Literature and Correspondence, box 17, CID.

81 E. F. McDonald to Hallowell Davis, 17 February 1950, RG032 Series 3: Hearing Aid Device Manufacturers, Literature and Correspondence, box 17, CID.

82 Responding to charges of negative advertising, E. F. McDonald Jr. wrote to Howard A. Carter of the AMA: "Poppycock! Nothing has shaken public confidence in the hearing aid industry so much as the fraudulent and/or misleading and merchandising practices which are described in 'Frauds and Facts.'" Nothing more than a complete cleanup of the industry, he continued, could rebuild public confidence; in the meantime, the industry should circulate the booklet as widely as possible to assure customers that their products were useful. Strangely,

he added in the letter that even Irving Schachtel supported *Frauds and Facts* and believed it would help Sonotone's business. E. F. McDonald to Howard A. Carter, 24 January 1952, RG032 Series 3: Hearing Aid Device Manufacturers, Literature and Correspondence, box 17, CID.

83 Hallowell Davis to O. D. Baker, 3 February 1954, HDP box 16, folder 16.

84 U.S. Senate Hearings, "Hearing Loss, Hearing Aids, and the Elderly," 156.

85 U.S. Senate Hearings, "Hearing Loss, Hearing Aids, and the Elderly," 2.

86 U.S. Senate Hearings, "Hearing Loss, Hearing Aids, and the Elderly," 11.

87 U.S. Senate Hearings, "Hearing Loss, Hearing Aids, and the Elderly," 78.

88 Hallowell Davis, "Rehabilitation of the Defective Hearing Patient," draft manuscript, 1952, "Current Therapy, 1952–1954," box 12, folder 12, HDP.

89 Roger Sanjek, *Gray Panthers* (Philadelphia: University of Pennsylvania Press, 2009), 35.

90 Sanjek, *Gray Panthers*, 37.

91 G. David Wallace, "Nader Study Report Blasts Ear Specialists, Agencies," *Albine Reporter-News* (30 September 1973), 11.

92 Sanjek, *Gray Panthers*, 36.

93 "Nader Blasts Hearing Aid Industry," *Washington Post* (30 September 1973), KBMA.

94 Monica Wilch, "Deaf to Criticism? Hearing Aid Industry Angered by Nader Report Calling Sales Practices 'Unacceptable,'" *The Wheeling Herald* (4 October 1973), 7.

95 Nancy L. Ross, "Hearing Aids: Who's Paying for What and Should They?" *Washington Post*, 18 November 1973.

96 "1973—A Challenge and an Opportunity," *Hearing Dealer* (January 1973).

97 Ross, "Hearing Aids."

98 "Unrelated to Dr. Hall's [Edwin P. Hall of FTC] specific questions, but surely related to his overall interest are the two advertisements from St. Louis newspapers enclosed. . . . To say the least, Audivox has performed quite a trick in delivering sounds '. . . right to your center of understanding.'" A. Goodman to Hallowell Davis, 26 September 1958, HDP box 23, folder 1.

99 U.S. Senate Hearings, "Hearing Aids and the Older American," Subcommittee on Consumer Interests of the Elderly, Special Committee on Aging, U.S. Senate, 93th Congress, 1st Session, 11 September 1973.

100 U.S. Senate Hearings, "Hearing Aids and the Older American," 218. Ruben added: "Another thing that has happened, I know in Toronto, Canada, by using aids from England, they cost the patient or the hospitals somewhere between $15 and $20 apiece. If the hearing aid does not work, they just take it off, throw it away, and give them another one if it breaks down. It is perfectly possible to mass produce these, if you mass produce them, you might cut your quality, but our quality could not be much worse than it is now" (222).

101 "Salesmen Deceive the Deaf," *American* (Odessa, Texas), 30 August 1977.

102 Ron E. Watkins, a certified member of the National Hearing Aid Society, for instance, responded to the FTC's failure to regulate the hearing aid industry: "It amazes me to find so much attention given to the hearing aid industry. It is a very tiny industry and does not justify the attention and expenditure given to it. It is almost as though there is a vendetta in progress. . . . What is the FTC doing in the future? The Food and Drug Administration issued their findings over a year ago

and gave the industry a clean bill of health. We are so burdened with bureaus, agencies, and commissions that one doesn't seem to know what the other is doing and doesn't seem to care. They are so high-handed they act tyrannical at times. We are being strangled by them." Newspaper clipping, *Daily News*, 2 February 1979, Gallaudet Archives.

103 Michael Isikoff, "Hearing Aids and the FTC," *Washington Post*, 7 June 1984.
104 U.S. Senate Hearings, "Hearing Loss, Hearing Aids, and the Elderly," 50.
105 Hallowell Davis to James M. MacMillian, 16 January 1961, HDP box 28, folder 4.

Epilogue

1 Leonard M. Elstad, "Message to Parents of Deaf Children," *Rochester Advocate* 69 (December 1948): 5.
2 Stuart Blume, *The Artificial Ear: Cochlear Implants and the Culture of Deafness* (New Brunswick, NJ: Rutgers University Press, 2010), 83.
3 Adrien A. Eshraghi et al., "The Cochlear Implant: Historical Aspects and Future Prospects," *Anatomical Record: Advances in Integrative Anatomy and Evolutionary Biology* 295 (2012): 1967–1980.
4 Albert Mudry and Mara Mills, "The Early History of the Cochlear Implant," *JAMA Otolaryngology—Head and Neck Surgery* 139.5 (May 2013): 446–453.
5 Lydia Denworth, *I Can Hear You Whisper: An Intimate Journey Through the Science of Sound and Language* (New York: Plume Books, 2014), 94.
6 Douglas Martin, "Dr. William F. House, Inventor of Pioneering Ear-Implant Device, Dies at 89," *New York Times*, 15 December 2012. http://www.nytimes.com /2012/12/16/health/dr-william-f-house-inventor-of-cochlear-implant-dies.html/.
7 "Success for the 'Bionic Ear,'" *Time*, 12 March 1984.
8 "Ear Implant Is Approved by the FDA: Deaf Can Hear Everyday Sounds," *Washington Post*, 30 November 1984, F1; Gerald E. Loeb, "The Functional Replacement of the Ear," *Scientific American* 252.2 (February 1985): 104–111.
9 "The Electric Ear," *Newsweek*, 1 April 1974, 50.
10 Harlan Lane, *The Mask of Benevolence: Disabling the Deaf Community* (New York: Alfred A. Knopf, 1992), 3.
11 Jane E. Brody, "For a Handful of the Deaf, a Medial Innovation That Breaks through the Wall of Silence," *New York Times*, 20 September 1990, B8.
12 Clare, *Brilliant Imperfection*, 9.
13 Laura Mauldin, *Made to Hear: Cochlear Implants and Raising Deaf Children* (Minneapolis: University of Minnesota Press, 2016).
14 Blume, *The Artificial Ear*, 174.
15 Gavin Francis, "The Mysterious World of the Deaf," *The New York Review*, 20 November 2014, 46.
16 Felicity Barringer, "Pride in a Soundless World: Deaf Oppose a Hearing Aid," *New York Times*, 16 May 1993, 1.
17 Bharat Jayram Venkat, "Cures," *Public Culture* 28.3 (2016): 494.
18 Barringer, "Pride in a Soundless World," 1.
19 Marie Arana-Ward, "As Technology Advances, a Bitter Debate Divides the Deaf," *Washington Post*, 11 May 1987, A01; Edward Dolnick, "Deafness as Culture," *Atlantic Monthly* (September 1993): 37–53.

20 Francis, "The Mysterious World of the Deaf."

21 "NAD Position Statement on Cochlear Implants (2000)," National Association of the Deaf, https://www.nad.org/about-us/position-statements/position-statement-on-cochlear-implants/.

22 Mauldin, *Made to Hear*, 9.

23 Elizabeth Engelman, "13 Reasons to Sign to Your Hard of Hearing or Cochlear Implanted Child," *On Deafness: Raising Deaf Children in a Hearing World*, 12 September 2016, https://ondeafness.com/2016/09/12/14-reason-to-sign-to-your-hard-of-hearing-or-cochlear-implanted-child/.

24 Graeme Gooday and Karen Sayer, *Managing the Experience of Hearing Loss in Britain, 1830–1930* (London: Palgrave Macmillan, 2017).

25 Andrew Solomon, *Far from the Tree: Parents, Children, and the Search for Identity* (New York: Scribner, 2012), 112.

26 Blume, *The Artificial Ear*, 8.

27 ODAS Newsletter (Winter 1968), Alexander Graham Bell Association for the Deaf: Oral Hearing 1983–1986. Clarke School Archives, UMass Amherst Special Collections MS 472 Clarke box 2, Series 1.

28 Sally Williams, "Why Not All Deaf People Want to Be Cured," *Telegraph*, 13 September 2012, http://www.telegraph.co.uk/culture/9526045/Why-not-all-deaf-people-want-to-be-cured.html/.

29 David N. F. Fairbanks, "'Miracle Cures' for Deafness: A Study of Unproven Remedies for Treating the Deaf," unpublished thesis for fellowship in the American Laryngological, Rhinological and Otological Society, 1975. Extracts from this manuscript are published in David N. F. Fairbanks, "Unproven Remedies for Treating the Deaf," *Ear, Nose, and Throat Journal* (November 1981): 71–78. Special thanks to Robert K. Jackler for providing me with a copy.

30 Clare, *Brilliant Imperfection*, 87.

31 Owen Wrigley, *The Politics of Deafness* (Washington, D.C.: Gallaudet University Press, 1997), 94.

32 Liza Mundy, "A World of Their Own," *Washington Post*, 31 March 2002, 22–43.

33 Christopher Krentz, *Writing Deafness: The Hearing Line in Nineteenth-Century American Literature* (Chapel Hill: University of North Carolina Press, 2007), 211.

34 Allegra Ringo, "Understanding Deafness: Not Everyone Wants to Be 'Fixed,'" *Atlantic*, 9 August 2013.

35 Earnest Elmo Calkins, *"And Hearing Not—" Annals of an Adman* (New York: Charles Scribner's Sons, 1946), 308.

36 Brenda Jo Brueggemann, *Lend Me Your Ear: Rhetorical Constructions of Deafness* (Washington, D.C.: Gallaudet University Press, 1999), 158.

37 Carol Padden and Tom Humphries, *Inside Deaf Culture* (Cambridge: Harvard University Press, 2005), 163.

38 Thomas G. Tickle, *Ears that Hear* (Boston: John Hancock Mutual Life Insurance Company, 1950), National Library of Medicine, W6 P3 v.7481 box 979 no. 22.

39 Solomon, *Far from the Tree*, 37.

Selected Bibliography

Abbott, Andrew. *The System of Professions: An Essay on the Division of Expert Labor*. Chicago: University of Chicago Press, 1988.

Agnew, Jeremy. *Medicine in the Old West: A History, 1850–1900*. Jefferson, NC: McFarland & Company Inc., 2010.

Akerlof, George A., and Robert J. Schiller. *Phishing for Phools: The Economics of Manipulation and Deception*. Princeton: Princeton University Press, 2015.

Applegate, Edd. *The Rise of Advertising in the United States: A History of Innovation to 1960*. Lanham, MD: Scarecrow Press, 2012.

Badaracco, Claire. *Prescribing Faith: Medicine, Media, and Religion in American Culture*. Waco, TX: Baylor University Press, 2007.

Baynton, Douglas C. *Defectives in the Land: Disability and Immigration in the Age of Eugenics*. Chicago: University of Chicago Press, 2016.

Baynton, Douglas C. *Forbidden Signs: American Culture and the Campaign Against Sign Language*. Chicago: University of Chicago Press, 1996.

Benedetti, Paul, and Wayne MacPhail. *Spin Doctors: The Chiropractic Industry Under Examination*. Toronto: Dundurn Press, 2002.

Berghoff, Harmutt, Phillip Scranton, and Uwe Spiekermann (eds.). *The Rise of Marketing Research*. New York: Palgrave Macmillan, 2012.

Berger, Kenneth W. *The Hearing Aid: Its Operation and Development*. Detroit: National Hearing Aid Society, 1970.

Bertucci, Paola. "The Shocking Bag: Medical Electricity in Mid-18th-Century London," *Nuova Voltiana: Studies on Volta and His Times* 5 (2003): 31–42.

Bertucci, Paola. "Therapeutic Attractions: Applications of Electricity to the Art of Healing." In H. A. Whitaker, C. U. M. Smith, and S. Finger (eds.), *Brain, Mind and Medicine: Essays in Eighteenth-Century Neuroscience* (Boston: Springer, 2007), 271–283.

Blume, Stuart. *The Artificial Ear: Cochlear Implants and the Culture of Deafness*. New Brunswick, NJ: Rutgers University Press, 2010.

Bouton, Katherine. *Shouting Won't Help: Why I — and 50 Million Americans — Can't Hear You*. New York: Picador, 2014.

Boyle, Eric. *Quack Medicine: A History of Combating Health Fraud in Twentieth-Century America*. Santa Barbara: Prager, 2013.

Branson, Jan, and Don Miller. *Damned for Their Difference: The Cultural Construction of Deaf People as Disabled*. Washington, D.C.: Gallaudet University Press, 2002.

Brueggemann, Brenda Jo. *Deaf Subjects: Between Identities and Places*. New York: New York University Press, 2009.

Brueggemann, Brenda Jo. *Lend Me Your Ear: Rhetorical Constructions of Deafness*. Washington, D.C.: Gallaudet University Press, 1999.

Brett, Dorothy. *Lawrence and Brett: A Friendship*. Santa Fe, NM: Sunstone Press, 2006.

Brune, Jeffrey A. and Daniel J. Wilson (eds.). *Disability and Passing: Blurring the Lines of Identity*. Philadelphia: Temple University Press, 2013.

Buchanan, Robert M. *Illusions of Equality: Deaf Americans in School and Factory, 1850–1950*. Washington, D.C.: Gallaudet University Press, 1999.

Burch, Susan. *Signs of Resistance: American Deaf Cultural History, 1900 to World War II*. New York: New York University Press, 2002.

Burch, Susan, and Hannah Joyner. *Unspeakable: The Story of Junius Wilson*. Chapel Hill: University of North Carolina Press, 2007.

Burney, Ian. *Bodies of Evidence: Medicine and the Politics of the English Inquest, 1839–1926*. Baltimore: Johns Hopkins University Press, 2000.

Burnham, John C. *Health Care in America: A History*. Baltimore: Johns Hopkins University Press, 2015.

Burnham, John C. *How Superstition Won and Science Lost: Popularizing Science and Health in the United States*. New Brunswick: Rutgers University Press, 1987.

Bynum, W. F. *Science and Practice of Medicine in the Nineteenth Century*. Cambridge: Cambridge University Press, 1994.

Carpenter, Mary Wilson. *Health, Medicine, and Society in Victorian England*. Santa Barbara: ABC-CLIO, 2010.

Carter, Julian B. *The Heart of Whiteness: Normal Sexuality and Race in America, 1880–1940*. Durham: Duke University Press, 2007.

Cayleff, Susan E. *Nature's Path: A History of Naturopathic Healing in America*. Baltimore: Johns Hopkins University Press, 2016.

Churman, Sarah. *Powered On: On the SOUNDS I Choose to Hear and the NOISE I Don't*. Pensacola: Indigo River Publishing, 2012.

Clare, Eli. *Brilliant Imperfection: Grappling with Cure*. Durham: Duke University Press, 2017.

Clarke, Adele E., Laura Mamo, Jennifer Ruth Fosket, Jennifer R. Fishman, and Janet K. Shim (eds.). *Biomedicalization: Technoscience, Health, and Illness in the U.S.* Durham: Duke University Press, 2010.

Cleall, Esme. "'Deaf to the Word': Gender, Deafness, and Protestantism in Nineteenth-Century Britain." *Gender and History* 25.3 (November 2013): 590–603.

Cones, Harold N., and John H. Bryant. *Zenith Radio: The Glory Years 1936–1945*. Atglen, PA: Schiffer Publishing Ltd., 2003.

Connor, J. T. H., and Felicity Pope. "A Shocking Business: The Technology and Practice of Electrotherapeutics in Canada, 1840s to 1940s." *Material Culture Review* 49 (Spring 1999): 60–70.

Conrad, Peter. *The Medicalization of Society: On the Transformation of Human Conditions into Treatable Disorders*. Baltimore: Johns Hopkins University Press, 2007.

Cook, James. *Remedies and Rackets: The Truth about Patent Medicines Today*. New York: W. W. Norton & Company, 1958.

Couser, G. Thomas. *Recovering Bodies: Illness, Disability, and Life Writing*. Madison: University of Wisconsin Press, 1997.

Covey, Herbert C. *African-American Slave Medicine: Herbal and Non-Herbal Treatments*. Plymouth: Lexington Books, 2007.

Cramp, Arthur J. *Nostrums and Quackery*. Chicago: Press of the American Medical Association, 1912.

Cramp, Arthur J. *Nostrums and Quackery*, vol. 2. Chicago: Press of the American Medical Association, 1921.

Cramp, Arthur J. *Nostrums and Quackery and Pseudo-Medicine*, vol. 3. Chicago: Press of the American Medical Association, 1936.

Creadick, Anna G. *Perfectly Average: The Pursuit of Normality in Postwar America*. Amherst: University of Massachusetts Press, 2010.

Cruikshank, Jeffrey L., and Arthur W. Schulz. *The Man Who Sold America: The Amazing (But True!) Story of Albert D. Lasker and the Creation of the Advertising Industry*. Boston: Harvard Business Press, 2010.

Cryle, Peter, and Elizabeth Stephens. *Normality: A Critical Genealogy*. Chicago: University of Chicago Press, 2017.

Curran, James R., and Jason R. Galster. "The Master Hearing Aid." *Trends in Amplification* 17.2 (2013): 108–134.

Davidson, Jonathan. *A Century of Homeopaths: Their Influence on Medicine and Health*. New York: Springer Books, 2014.

Davis, Hallowell. *Hearing and Deafness: A Guide for Laymen*. New York: Reinhart & Company, Inc., 1954.

Davis, Lennard J. (ed.). *The Disability Studies Reader*. 4th ed. New York: Routledge, 2013.

Davis, Lennard J. *The End of Normal: Identity in a Biocultural Era*. Ann Arbor: University of Michigan Press, 2013.

Davis, Lennard J. *Enforcing Normalcy: Disability, Deafness, and the Body*. New York: Verso Books, 1995.

De la Peña, Carolyn Thomas. *The Body Electric: How Strange Machines Built the Modern American*. New York: New York University Press, 2003.

Denworth, Lydia. *I Can Hear You Whisper: An Intimate Journey Through the Science of Sound and Language*. New York: Plume Books, 2014.

Edwards, R. A. R. *Words Made Flesh: Nineteenth-Century Deaf Education and the Growth of Deaf Culture*. New York: New York University Press, 2012.

Esmail, Jennifer. *Reading Victorian Deafness: Signs and Sounds in Victorian Literature and Culture*. Athens: Ohio University Press, 2013.

Fairbanks, David N. F. "Unproved Remedies for Treating the Deaf." *Ear, Nose, and Throat Journal* 60 (November 1981): 71–87.

Ferry, A. P. "'Professor' William C. Wilson and His Actina Electric Power Battery for Curing Ocular Diseases." *Ophthalmology* 105.2 (1998): 238–248.

Folk, Holly. *The Religion of Chiropractic Populist Healing from the American Heartland*. Chapel Hill: University of North Carolina Press 2017.

Gannon, Jack R. *Deaf Heritage: A Narrative History of Deaf America*. Revised and expanded ed. Washington, D.C.: Gallaudet University Press, 2012.

Goldstein, Max A. *Problems of the Deaf*. St. Louis: Laryngoscope Press, 1933.

Goffman, Erving. *Stigma: Notes on the Management of Spoiled Identity*. New York: Simon & Schuster, Inc., 1963.

Gooday, Graeme, and Karen Sayer. *Managing the Experience of Hearing Loss in Britain, 1830–1930*. London: Palgrave Macmillan, 2017.

Gooday, Graeme, and Karen Sayer. "Purchase, Use, and Adaptation: Interpreting 'Patented' Aids to the Deaf in Victorian Britain." In Claire Jones (ed.)., *Rethinking Modern Prostheses in Anglo-American Commodity Cultures, 1829–1939*. Manchester: Manchester University Press. 27–47.

Gregory, Susan, and Gillian M. Hartley (eds.). *Constructing Deafness*. Trowbridge, Wiltshire: Redwood Books, 1991.

Harmon, Kristen C. "Growing Up to Become Hearing: Dreams of Passing in Oral Deaf Education." In Jeffrey A. Brune and Daniel J. Wilson (eds.), *Disability and Passing: Blurring the Lines of Identity*. Philadelphia: Temple University Press, 2013. 167–198.

Harrison, Andrew. *The Life of D. H. Lawrence*. Chichester, West Sussex: Wiley Blackwell, 2016.

Hatfield, Gabrielle. *Encyclopedia of Folk Medicine: Old World and New Word Traditions*. Santa Barbara: ABC-CLIO, 2004.

Heiner, Marie Hays. *Hearing Is Believing*. Cleveland: World Publishing Co., 1949.

Hetchlinger, Adelaide. *The Great Patent Medicine Era, or, Without Benefit of Doctor*. New York: Grosset & Dunlap Inc., 1970.

Hignett, Sean. *Brett: From Bloomsbury to New Mexico—A Biography*. London: Hodder and Stoughton, 1984.

Hoffman, Frank, and Martin Manning. *Herbal Medicine and Botanical Medical Fads*. New York: Routledge, 2012.

Hogan, Neal C. *Unhealed Wounds: Medical Malpractice in the Twentieth Century*. New York: LFB Scholarly Publishing, 2003.

Holmes, Robert Wendell, III. "Substance of the Sun: The Cultural History of Radium Medicines in America." Ph.D. thesis, University of Texas at Austin, 2010.

Illich, Ivan. *Medical Nemesis: The Expropriation of Health*. London: Calder & Boyars, 1975.

Janik, Erika. *Marketplace of the Marvelous: The Strange Origins of Modern Medicine*. Boston: Beacon Press, 2014.

Jones, Claire L. (ed.). *Rethinking Modern Prostheses in Anglo-American Commodity Cultures, 1820–1939*. Manchester: Manchester University Press, 2017.

Joyner, Hannah. *From Pity to Pride: Growing Up Deaf in the Old South*. Washington, D.C.: Gallaudet University Press, 2004.

Juhnke, Eric S. *Quacks and Crusaders: The Fabulous Careers of John Brinkley, Norman Baker, and Harry Hoxsey*. Lawrence: University Press of Kansas, 2002.

Keeler, Harry Stephen. *The Man with the Magic Eardrums*. Vancleave, MS: Ramble House, 2010; originally published New York: Dutton, 1939.

Klawiter, Maren. *The Biopolitics of Breast Cancer: Changing Cultures of Disease and Activism*. Minneapolis: University of Minnesota Press, 2008.

Krentz, Christopher. *Writing Deafness: The Hearing Line in Nineteenth-Century American Literature*. Chapel Hill: University of North Carolina Press, 2007.

Ladd, Paddy. *Understanding Deaf Culture: In Search of Deafhood*. Clevedon: Multilingual Matters, 2003.

Lane, Harlan. *The Mask of Benevolence: Disabling the Deaf Community*. New York: Alfred A. Knopf, 1992.

Lang, Harry G., and Bonnie Meaath-Lang. *Deaf Persons in the Arts and Sciences: A Biographical Dictionary*. Westport, CT: Greenwood Press, 1995.

Leach, William. *Land of Desire: Merchants, Power, and the Rise of a New American Culture*. New York: Vintage Books, 1993.

Lejaq, Seth Stein. "The Bounds of Domestic Healing: Medical Recipes, Storytelling, and Surgery in Early Modern England." *Social History of Medicine* 26.3 (2013): 452–468.

Lerner, Baron H. *The Breast Cancer Wars: Hope, Fear, and the Pursuit of a Cure in Twentieth-Century America*. New York: Oxford University Press, 2001.

Lindenov, Harland. *The Etiology of Deaf-Mutism with Special Reference to Heredity*. Copenhagen: Einar Munksgaard, 1945.

Loeb, Lori. "Consumerism and Commercial Electrotherapy: The Medical

Battery Company in Nineteenth-Century London." *Journal of Victorian Culture* 4:2 (1999): 252–275.

Loeb, Lori. *Consuming Angels: Advertising and Victorian Women*. Oxford: Oxford University Press, 1994.

Longmore, Paul K., and Lauri Umansky (eds.). *The New Disability History: American Perspectives*. New York: New York University Press, 2001.

Mansfield, Katherine. *The Collected Letters of Katherine Mansfield*. Vol. 4, *1920–1921*. Vincent O'Sullivan and Margaret Scott (eds.). Oxford: Clarendon Press, 1996.

Marchand, Roland. *Advertising the American Dream: Making Way for Modernity 1920–1940*. Berkeley: University of California Press, 1985.

Markides, Andreas. "Some Unusual Cures for Deafness." *Journal of Laryngology and Otology* 96.6 (June 1982): 479–490.

Mauldin, Laura. *Made to Hear: Cochlear Implants and Raising Deaf Children*. Minneapolis: University of Minnesota Press, 2016.

McDonald, Donna. *The Art of Being Deaf: A Memoir*. Washington, D.C.: Gallaudet University Press, 2014.

McDonald, Donna. "Hearsay: How Stories About Deafness and Deaf People Are Told." Ph.D. thesis, University of Queensland, 2010.

Mills, Mara. "Cochlear Implants after Fifty Years: A History and an Interview with Charles Graser." In Sumanth Gopinath and Jason Stanyek (eds.), *The Oxford Handbook of Mobile Music Studies*, vol. 1. Oxford: Oxford University Press, 2014. 261–297.

Mills, Mara. "Hearing Aids and the History of Electronics Miniaturization." *IEEE Annals of the History of Computing* 33.2 (April–June 2011): 24–45.

Mills, Mara. "When Mobile Communication Technologies Were New." *Endeavor* 33.4 (2009): 140–146.

Morus, Iwan Rhys. "Bodily Disciplines and Disciplined Bodies: Instruments, Skills, and Victorian Electrotherapeutics." *Social History of Medicine* 19.2 (2006): 241–259.

Morus, Iwan Rhys. "Marketing the Machine: The Construction of Electrotherapeutics as Viable Medicine in Early Victorian England." *Medical History* 36 (1992): 34–52.

Morus, Iwan Rhys. *Shocking Bodies: Life, Death, and Electricity in Victorian England*. Stroud: History Press, 2011.

Mounsey, Chris (Ed.). *The Idea of Disability in the Eighteenth Century*. Lewisburg: Bucknell University Press, 2014.

Nielsen, Kim E. *A Disability History of the United States*. Boston: Beacon Press, 2012.

Nye, David. *Electrifying America: Social Meanings of a New Technology, 1880–1940*. Cambridge: MIT Press, 1990.

Odgen, Emily. *Credulity: A Cultural History of US Mesmerism*. Chicago: University of Chicago Press, 2018.

Ott, Katherine. "Prosthetics." In Rachel Adams, Benjamin Reiss, and David Serlin (eds.), *Keywords for Disability Studies*. New York: New York University Press, 2015.

Ott, Katherine, David Serlin, and Stephen Mihm (eds.). *Artificial Parts, Practical Lives: Modern Histories of Prosthetics*. New York: New York University Press, 2002.

Owen, Thomas McAdory. *History of Alabama and Dictionary of Alabama Biography*. Vol. 3. Chicago: S. J. Clarke Publishing Company, 1921.

Padden, Carol, and Tom Humphries. *Inside Deaf Culture*. Cambridge: Harvard University Press, 2005.

Paris, Joel. *Fads and Fallacies in Psychiatry*. London: Royal College of Physicians, 2013.

Porter, Roy. *Bodily Politic: Disease, Death and Doctors in Britain, 1650–1900*. Ithaca, NY: Cornell University Press, 2001.

Porter, Roy. *Quacks: Fakers & Charlatans in Medicine*. Gloucestershire: Tempus Publishing Ltd., 1989.

Quartararo, Anne T. *Deaf Identity and Social Images in Nineteenth-Century France*. Washington, D.C.: Gallaudet University Press, 2008.

Reagan, Leslie J. *Dangerous Pregnancies: Mothers, Disabilities, and Abortion in Modern America*. Berkeley: University of California Press, 2010.

Rée, Jonathan. *I See a Voice: Deafness, Language, and the Senses—A Philosophical History*. New York: Metropolitan Books, 1999.

Rose, Sarah F. *No Right to Be Idle: The Invention of Disability, 1840s–1930s*. Chapel Hill: University of North Carolina Press, 2017.

Rosner, Lisa. "The Professional Context of Electrotherapeutics." *Journal of the History of Medicine and Allied Sciences* 43 (1988): 64–82.

Ross, Liz, Phil Lyon, and Craid Cathcart. "Pills, Potions, and Devices: Treatments for Hearing Loss Advertised in Mid-Nineteenth Century British Newspapers." *Social History of Medicine* 27.3 (2014): 530–556.

Rutkow, Ira. *Seeking the Cure: A History of Medicine in America*. New York: Scribner, 2010.

Sanchez, Rebecca. *Deafening Modernism: Embodied Language and Visual Poetics in American Literature*. New York: New York University Press, 2015.

Sanjek, Roger. *Gray Panthers*. Philadelphia: University of Philadelphia Press, 2009.

Schlich, Thomas (ed.). *The Palgrave Handbook of the History of Surgery*. London: Palgrave Macmillan, 2016.

Schudson, Michael. *The Good Citizen: A History of American Civic Life*. New York: Simon & Schuster, 1998.

Schuster, David G. *Neurasthenic Nation: America's Search for Health, Happiness, and Comfort, 1869–1920*. New Brunswick, NJ: Rutgers University Press, 2011.

Schwartz, Hillel. "Hearing Aids: Sweet Nothings, or an Ear for an Ear." In

Pat Kirkham (ed.), *The Gendered Object*. Manchester: Manchester University Press, 1996. 43–59.

Schwartz, Hillel. *Making Noise: From Babel to the Big Bang and Beyond*. Cambridge, MA: MIT Press, 2011.

Serlin, David. *Replaceable You: Engineering the Body in Postwar America*. Chicago: University of Chicago Press, 2004.

Shattuck, Roger. *The Forbidden Experiment: The Story of the Wild Boy of Aveyron*. New York: Kodansha International, 1980.

Shaw, Courtney Q. *Sex Ed., Segregated: The Quest for Sexual Knowledge in Progressive-Era America*. Rochester: University of Rochester Press, 2015.

Simmons, Michael. *Hearing Loss: From Stigma to Strategy*. London: Peter Owen Publishers, 2005.

Sivulka, Juliann. *Soap, Sex, and Cigarettes: A Cultural History of American Advertising*. 2nd ed. Boston: Wadsworth, 2012.

Solomon, Andrew. *Far from the Tree: Parents, Children, and the Search for Identity*. New York: Scribner, 2012.

Stark, James F. "'Recharge My Exhausted Batteries': Overbeck's Rejuvenator, Patenting, and Public Medical Consumers, 1924–37." *Medical History* 58.4 (2014): 498–518.

Stearns, Peter N. *Fat History: Bodies and Beauty in the Modern West*. New York: New York University Press, 2002.

Sterne, Jonathan. *The Audible Past: Cultural Origins of Sound Reproduction*. Durham: London: Duke University Press, 2003.

Stevens, Rosemary. *American Medicine and the Public Interest*. Berkeley: University of California Press, 1998.

Swamy, Ravi S., and Robert K. Jackler. "The Fickle Finger of Quackery in Otology: The Saga of Curtis H. Muncie, Osteopath." *Otology and Neurotology* 31 (2010): 846–855.

Thompson, Emily. *The Soundscape of Modernity: Architectural Acoustics and the Culture of Listening in America, 1900–1933*. Cambridge: MIT Press, 2002.

Thomson, Rosemarie Garland. *Extraordinary Bodies: Figuring Physical Disability in American Culture and Literature*. New York: Columbia University Press, 1997.

Tomes, Nancy. "Merchants of Health: Medicine and Consumer Culture in the United States, 1900–1940." *Journal of American History* 88.2 (2001): 519–547.

Tomes, Nancy. *Remaking the American Patient: How Madison Avenue and Modern Medicine Turned Patients into Consumers*. Chapel Hill: University of North Carolina Press, 2016.

Turner, David M., and Alun Withey. "Technologies of the Body: Polite Consumption and the Correction of Deformity in Eighteenth-Century England." *History* 99 (2014): 775–796.

Ueyama, Takahiro. *Health in the Marketplace: Professionalism, Therapeutic Desires, and Medical Commodification in Late-Victorian London.* Palo Alto, CA: Society for the Promotion of Science and Scholarship, 2010.

Van Cleve, John V. and Barry A. Crouch. *A Place of Their Own: Creating the Deaf Community in America.* Washington, D.C.: Gallaudet University Press, 1989.

Venkat, Bharat Jayram. "Cures." *Public Culture* 28.3 (2016): 475-497.

Virdi, Jaipreet [as Virdi-Dhesi]. "Curtis's Cephaloscope: Deafness and the Making of Surgical Authority in London, 1816-1845." *Bulletin of the History of Medicine* 87.5 (2013): 347-377.

Waits, Robert K. *The Medical Electricians: Dr. Scott and His Victorian Cohorts in Quackery.* Sunnyvale, CA: J.IV.IX Publications, 2013.

Weir, Neil. *Otolaryngology: An Illustrated History.* London: Butterworths, 1990.

Weiss, Harry B., and Howard R. Kemble. *The Great American Water-Cure Craze: A History of Hydropathy in the United States.* Trenton, NJ: Past Times Press, 1967.

Wexler, Anna. "The Medical Battery in the United States (1870-1920): Electrotherapy at Home and in the Clinic." *Journal of the History of Medicine and Allied Sciences* 72.2 (April 2017): 166-192.

Whorton, James C. *Nature Cures: The History of Alternative Medicine in America.* Oxford: Oxford University Press, 2002.

Wilson, Daniel J. "Passing in the Shadow of FDR: Polio Survivors, Passing, and the Negotiation of Disability." In Jeffrey A. Brune and Daniel J. Wilson (eds.), *Disability and Passing: Blurring the Lines of Identity,* 13-35. Philadelphia: Temple University Press, 2013.

Winzer, Margaret A. *From Integration to Inclusion: A History of Special Education in the 20th Century.* Washington, D.C.: Gallaudet University Press, 2009.

Woloshyn, Tania Anne. *Soaking Up the Rays of Light: Light Therapy and Visual Culture in Britain, c. 1890-1940.* Manchester: University of Manchester Press, 2017.

Wright, David. *Deafness.* New York: Stein and Day, 1969.

Wrigley, Owen. *The Politics of Deafness.* Washington, D.C.: Gallaudet University Press, 1997.

Yellon, Evan. *Surdus in Search of His Hearing: An Exposure of Aural Quacks and a Guide to Genuine Treatments and Remedies, Electric Aids, Lip-Reading and Employments for the Deaf Etc., Etc.* London: The Celtic Press, 1906.

Yellon, Evan. *Surdus in Search of His Hearing: An Exposure of Deafness Quacks, Frauds on the Deaf, and a Reliable Guide to the Best Means of Help for the Deaf.* London: Evan McLeod, 1910.

Young, James Harvey. *American Health Quackery.* Princeton: Princeton University Press, 1992.

Young, James Harvey. *The Medical Messiahs: A Social History of Health Quackery in Twentieth-Century America.* Princeton: Princeton University Press, 1992.

Young, James Harvey. *The Toadstool Millionaires: A Social History of Patent Medicines in America Before Federal Regulation.* Princeton: Princeton University Press, 1961.

Zola, Irving Kenneth. *Missing Pieces: A Chronicle of Living with a Disability.* Philadelphia: Temple University Press, 1982.

Index

Italicized page numbers indicate images.